Geographical Ecology

Geographical Ecology

Patterns in the Distribution of Species

Robert H. MacArthur

Princeton University Press
Princeton, New Jersey

Published by Princeton University Press, 41 William Street,
Princeton, New Jersey 08540
In the United Kingdom: Princeton University Press,
Guildford, Surrey

First Princeton Paperback printing, 1984

LCC 83-24477
ISBN 0-691-08353-3
ISBN 0-691-02382-4 (pbk.)

Reprinted by arrangement with Harper & Row, Publishers, Inc.

Clothbound editions of Princeton University Press books
are printed on acid-free paper, and binding materials are
chosen for strength and durability. Paperbacks, while
satisfactory for personal collections, are not usually suitable
for library rebinding.

Printed in the United States of America
by Princeton University Press,
Princeton, New Jersey

For
Betsy
Duncan, Alan, Lizzie, and Donny

Contents

Preface

In this book I have written about my favorite part of ecology, the part that combines the adventure of field work in varied places with the discipline of making nontrivial theory. It is the part best done by field naturalists, but naturalists who know what science is all about. Certainly one of the most successful areas of ecology, it is the part in which the comparisons that are intrinsic to all scientific effort are geographical ones.

Such a book could perhaps be based on a logical structure or sequence, but I have chosen a different arrangement, one that is more faithful to the actual spirit of the research. If, for example, one is going to compare an island with the mainland, one has to know some geology, some systematics, a good deal of evolution, and a great deal of ecology. But separate chapters on each of these would be misleading. Instead, a phenomenological arrangement, with one chapter on islands, one on the tropics, and so on, gives a more authentic picture of the subject. Of course, a logical outline might involve less redundancy, but since no one has ever accused me of excess redundancy in my writing, I do not fear the slight repetition or cross-referencing of the approach I have chosen. Furthermore, the references I have given have been included not just to document facts but as suggestions for further reading; and since most of the references are themselves studies of particular situations, they are, like the book, phenomenological.

My own lectures at Princeton University cover this material in one semester. In the book, I have separated the mathematical portions as appendixes to the chapters and in this way have given the teacher two options. Since the chapter texts are fully coherent without the mathematics, the teacher who prefers to emphasize the field studies may require some of the references, or other articles, as outside reading and also use the book for a one-semester course. Most of the mathematics in the appendixes should be understandable to anyone with a knowledge of calculus. In a few places elementary linear algebra is used, and in one or two, some probability theory. For students who have this preparation, much greater depth can be achieved. The student who both learns the mathematics and reads from the bibliography, to complete a year's course, is ready to do research on his own.

I wrote the book while convalescing in Vermont with no access to libraries, entirely from memory. The bibliography therefore contains only those references that have influenced me, and it is not an exhaustive list. I apologize to the authors of those equally good articles not included.

Nearly all the ecologists I know have influenced my thinking in one way or another, but four must be singled out, since this book would be unrecognizably different, or would be nonexistent, without them. G. Evelyn Hutchinson has for 20 years encouraged me to believe that my ideas are worthwhile and has provided many himself. Richard Levins has shared theoretical ideas with me on such a give-and-take basis that I could not possibly trace the history of many that are included here. E. O. Wilson showed me how interesting biogeography could be and wrote the lion's share of a joint book on islands. Jared Diamond rekindled my interest in geographical ecology by bringing to my attention how many ideas could be tested with field data. My debt to these four friends is very great.

E. O. Wilson, Jared Diamond, and Monte Lloyd provided invaluable further assistance by reading the manuscript and making many suggestions, almost all of which I accepted. My wife, Betsy, and my brother, John, read and improved selected chapters. Because of my remoteness from libraries, I requested reprints and unpublished material from various friends, all of whom responded promptly and with generosity. For these favors I am indebted to James Brown, Joseph Connell, Jared Diamond, Madhav Gadgil, William Heed, James Karr, Gordon Lark, Gordon Orians, Harry Recher, Howard Sanders, Thomas Schoener, Christopher Smith, Neal Smith, Otto Solbrig, John Terborgh, Ernest Williams, and E. O. Wilson.

My wife typed much of the manuscript. Grace Russell prepared illustrations, checked the manuscript, and handled permissions requests.

Mrs. Amy Kramer, Harper & Row editor, has been so interested in the book that she would have gone on making me improve its clarity indefinitely had she not already achieved a very rapid publication schedule. I greatly appreciate her help.

Robert H. MacArthur

Acknowledgments

Thanks are extended to the following authors and publishers for permission to quote copyrighted material in figures and tables, as listed below.

Figures 1–2 United States Department of Agriculture, *Climate and Man,* Washington, D.C., 1941, p. 703.

1–3 H. Flohn, *Climate and Weather,* World University Library, McGraw-Hill, New York, 1969, Figure 10, p. 47. Copyright 1969 by McGraw-Hill. Used with permission of McGraw-Hill Book Company.

1–5 H. Flohn, *Climate and Weather,* World University Library, McGraw-Hill, New York, 1969, Figure 34, p. 85. Copyright 1969 by McGraw-Hill. Used with permission of McGraw-Hill Book Company.

1–6 R. MacArthur and J. Connell, *The Biology of Populations,* Wiley, New York, 1966, Figure 1-10, p. 20.

1–7 R. MacArthur and J. Connell, *The Biology of Populations,* Wiley, New York, 1966, Figure 1-11, p. 21.

2–1 R. MacArthur and J. Connell, *The Biology of Populations,* Wiley, New York, 1966, Figure 6-1, p. 146.

2–2 M. Way, *Bull. Entomol. Res.* 44: 669–691, 1953, Figure 6, p. 678.

2–11 R. MacArthur, "The Theory of the Niche," in *Population Biology and Evolution,* R. C. Lewontin (ed.), Syracuse University Press, Syracuse, N. Y., 1968, Figure 2, p. 167.

3–1 T. W. Schoener and D. Janzen, "Notes on Environmental Determinants of Tropical Versus Temperate Insect Size Patterns," *Amer. Natur.* 102: 207–224, 1968, Figure 1, p. 216. Copyright © 1968 by The University of Chicago Press.

3–2 R. MacArthur and E. O. Wilson, *The Theory of Island*

Biogeography, Princeton University Press, Princeton, N.J., 1967, Figure 33, p. 108. Copyright © 1967 by Princeton University Press.

3–5 H. Hespenheide, "Food Preference and the Extent of Overlap in Some Insectivorous Birds, with Special Reference to Tyrannidae," *Ibis* 113: 59–72, 1971, Figure 1, p. 60.

3–6 H. Hespenheide, "Food Preference and the Extent of Overlap in Some Insectivorous Birds, with Special Reference to Tyrannidae," *Ibis* 113: 59–72, 1971, Figure 6, p. 66.

3–7 T. W. Schoener, "Optimal Size and Specialization in Constant and Fluctuating Environments: An Energy-Time Approach," in "Diversity and Stability in Ecological Systems," *Brookhaven Symp. Biol.* 22, 1969, Figure 5, p. 110.

5–2 D. S. Simberloff and E. O. Wilson, "Experimental Zoogeography of Islands: A Two-Year Record of Colonization," *Ecology* 51: 934–937, 1970, Figure 1, p. 936. Copyright by the Duke University Press.

5–3 D. S. Simberloff and E. O. Wilson, "Experimental Zoogeography of Islands: The Colonization of Empty Islands," *Ecology* 50: 278–296, 1969, Appendix 6, p. 295. Copyright 1969 by the Duke University Press.

5–6 E. O. Wilson, "Adaptive Shift and Dispersal in a Tropical Ant Fauna," *Evolution* 13: 122–144, 1959, Figure 9, p. 137.

5–7 E. O. Wilson, "Adaptive Shift and Dispersal in a Tropical Ant Fauna," *Evolution* 13: 122–144, 1959, Figure 10, p. 137.

5–8 R. MacArthur and E. O. Wilson, *The Theory of Island Biogeography,* Princeton University Press, Princeton, N.J., 1967, Figures 27A and 27B, pp. 74–75. Copyright © 1967 by Princeton University Press.

5–9 S. D. Webb, "Extinction-Origination Equilibria in Late Cenozoic Land Mammals of North America," *Evolution* 23: 688–702, 1969, Figure 1, p. 689.

5–16 E. O. Wilson and R. W. Taylor, "An Estimate of the Potential Evolutionary Increase in Species Diversity in the Polynesian Ant Fauna," *Evolution* 21: 1–10, 1967, Figure 3, p. 5.

5–17 R. MacArthur and E. O. Wilson, *The Theory of Island Biogeography,* Princeton University Press, Princeton, N.J., 1967, Figure 2, p. 8. Copyright © 1967 by Princeton University Press.

5–18 F. Viulleumier, "Insular Biogeography in Continental Regions: The Northern Andes of South America," *Amer. Natur.* 104: 373–388, 1970, Figure 1, p. 374. Copyright © 1970 by The University of Chicago Press.

5–21 T. W. Schoener, "The Ecological Significance of Sexual Dimorphism in Size in the Lizard *Anolis conspersus,*" *Science* 155: 474–477, 1967, Figure 3, p. 477. Copyright 1967 by the American Association for the Advancement of Science.

5–22 D. Lack, *Ecological Isolation in Birds,* Blackwell, Oxford, 1971, Figure 53, p. 231.

6–1 J. R. Hastings and R. M. Turner, *The Changing Mile,* The University of Arizona Press, Tucson, 1965, Figure 5, p. 20. Copyright 1965 by the Arizona Board of Regents.

6–2 W. H. Pearsall, *Mountains and Moorlands,* Collins, London, 1950, Figure 15, p. 50.

6–3 J. Terborgh, "Distribution on Environmental Gradients: Theory and a Preliminary Interpretation of Distributional Patterns in the Avifauna of the Cordillera Vilcabamba, Peru," *Ecology* 52: 23–40, 1971, Figure 9, p. 33. Copyright 1971 by the Duke University Press.

6–4 J. Terborgh, "Distribution on Environmental Gradients: Theory and a Preliminary Interpretation of Distributional Patterns in the Avifauna of the Cordillera Vilcabamba, Peru," *Ecology* 52: 23–30, 1971, Figure 12, p. 36. Copyright 1971 by the Duke University Press.

6–5 J. M. Diamond, "Ecological Consequences of Island Colonization by Southwest Pacific Birds, I: Types of

Niche Shifts," *Nat. Acad. Sci., Proc.* 67: 529–536, 1970, Figure 1, p. 531.

6–7 C. Smith, "The Coevolution of Pine Squirrels (*Tamiasciurus*) and Conifers," *Ecol. Monogr.* 40(3): 349–371, 1970, Figure 4, p. 354. Copyright 1970 by the Duke University Press.

6–9 C. H. Lowe, W. B. Heed, and E. A. Halpern, "Supercooling of the Saguaro Species *Drosophila nigrospiracula* in the Sonoran Desert," *Ecology* 48: 984–985, 1967, Figure 1, p. 985. Copyright 1967 by the Duke University Press.

6–10 W. A. Niering, R. H. Whittaker, and C. H. Lowe, "The Saguaro: A Population in Relation to Environment," *Science* 142: 15–23, 1963, Figure 1, p. 17. Copyright 1963 by the American Association for the Advancement of Science.

6–11 C. Robbins and W. T. Van Velzen, "Breeding Bird Survey 1967 and 1968," Bureau of Sport Fisheries and Wildlife, Special Scientific Report 124, Washington, D.C., 1969, Figures 33 and 34, pp. 62–63.

6–14 D. Lack, *Ecological Isolation in Birds,* Blackwell, Oxford, 1971, Figure 6, p. 34. Drawn by Robert Gillmor.

6–15 H. F. Recher, "Bird Species Diversity and Habitat Diversity in Australia and North America," *Amer. Natur.* 103: 75–80, 1969, Figure 1, p. 77. Copyright © 1969 by The University of Chicago Press.

6–16 D. Lack, *Ecological Isolation in Birds,* Blackwell, Oxford, 1971, Figure 12, p. 51. Drawn by Robert Gillmor.

6–17 M. L. Cody, "On the Methods of Resource Division in Grassland Bird Communities," *Amer. Natur.* 102: 107–147, 1968, Figure 11, p. 135. Copyright © 1968 by The University of Chicago Press.

6–18 L. R. Holdridge, *Life Zone Ecology,* Tropical Science Center, San José, Costa Rica, 1967, Figure 1, pp. 16–17.

6–19 R. H. Whittaker, "Evolution of Diversity in Plant Communities" in "Diversity and Stability in Ecological Systems," *Brookhaven Symp. Biol.* 22, 1969, Figures 5 and 6, pp. 183–184.

7–4 H. L. Sanders, "Benthic Marine Diversity and the Stability Time Hypothesis," in "Diversity and Stability in Ecological Systems," *Brookhaven Symp. Biol.* 22, 1969, Figure 6, p. 78.

7–5 H. L. Sanders, "Benthic Marine Diversity and the Stability Time Hypothesis," in "Diversity and Stability in Ecological Systems," *Brookhaven Symp. Biol.* 22, 1969, Figure 7, p. 79.

7–8 R. MacArthur, "Patterns of Species Diversity," *Biol. Rev.* 40: 510–533, 1965, Figure 5, p. 524. Copyright 1965 by Cambridge University Press.

7–9 R. T. Paine, "Food Web Complexity and Species Diversity," *Amer. Natur.* 100: 65–76, 1966, Figures 2 and 3, pp. 68–69. Copyright © 1966 by The University of Chicago Press.

8–2 D. W. Snow, "A Field Study of the Black and White Manakin (*Manacus manacus*) in Trinidad," *Zoologica* 47: 65–104, 1962, Figure 18, p. 95.

8–3 N. Smythe, "Neotropical Fruiting Seasons and Seed Dispersal," *Amer. Natur.* 104: 25–35, 1970, Figure 3, p. 28. Copyright © 1970 by The University of Chicago Press.

8–4 R. MacArthur, "On the Breeding Distribution Pattern of North American Migrant Birds," *Auk* 76: 318–325, 1959, Figure 1, p. 319.

8–5 T. W. Schoener, "Large-Billed Insectivorous Birds: A Precipitous Diversity Gradient," *Condor* 73: 154–161, 1971, Figure 1, p. 155.

8–6 T. W. Schoener, "Large-Billed Insectivorous Birds: A Precipitous Diversity Gradient," *Condor* 73: 154–161, 1971, Figure 3, p. 158.

8–8 R. MacArthur, "Patterns of Communities in the Tropics," *Biol. J. Linnaean Soc.* 1: 19–30, 1969, Figure 1.

8–9 A. Fischer, "Latitudinal Variations in Organic Diversity," *Evolution* 14: 64–81, 1960, Figure 12, p. 70.

8–10 F. G. Stehli, "Taxonomic Diversity Gradients in Pole Location: The Recent Model," in *Evolution and Environment,* Peabody Museum Centennial Symposium, Yale University Press, New Haven, Conn., 1968, Figure 20, p. 192.

8–11 J. Kikkawa and W. T. Williams, "Altitudinal Distribution of Land Birds in New Guinea," *Search* 2: 64–69, 1971, Figure 1, p. 65.

8–12 R. MacArthur, "Patterns of Communities in the Tropics," *Biol. J. Linnaean Soc.* 1: 19–30, 1969, Figure 2.

8–13 H. L. Sanders, "Marine Benthic Diversity: A Comparative Study," *Amer. Natur.* 102: 243–282, 1968, Figure 3, p. 250. Copyright © 1968 by The University of Chicago Press.

8–14 G. G. Simpson, *The Geography of Evolution,* Chilton, Philadelphia, 1965, Figure 22, p. 107.

9–1 M. B. Davis, "Palynology and Environmental History During The Quartenary Period," *Amer. Sci.* 57(3): 317–332, 1969, Figure 4, p. 321.

9–6 P. J. Darlington, *Zoogeography,* Wiley, New York, 1957, Figure 46, p. 427.

Tables 5–1 J. M. Diamond, "Avifaunal Equilibria and Species Turnover Rates on the Channel Islands of California," *Nat. Acad. Sci., Proc.* 64: 57–63, 1969, Table 1, p. 59.

8–1 N. Kusnezov, "Numbers of Species of Ants in Faunae of Different Latitudes," *Evolution* 11: 298–299, 1957.

8–2 R. Patrick, "The Cathewood Foundation Peruvian Amazon Expedition," Monographs of the Academy of National Sciences of Philadelphia, 1966, Table 13, p. 35.

Introduction

To do science is to search for repeated patterns, not simply to accumulate facts, and to do the science of geographical ecology is to search for patterns of plant and animal life that can be put on a map. The person best equipped to do this is the naturalist who loves to note changes in bird life up a mountainside, or changes in plant life from mainland to island, or changes in butterflies from temperate to tropics. But not all naturalists want to do science; many take refuge in nature's complexity as a justification to oppose any search for patterns. This book is addressed to those who do wish to do science. Doing science is not such a barrier to feeling or such a dehumanizing influence as is often made out. It does not take the beauty from nature. The only rules of scientific method are honest observations and accurate logic. To be great science it must also be guided by a judgment, almost an instinct, for what is worth studying. No one should feel that honesty and accuracy guided by imagination have any power to take away nature's beauty.

Science should be general in its principles. A well-known ecologist remarked that any pattern visible in my birds but not in his Paramecium *would not be interesting, because, I presume, he felt it would not be general. The theme running through this book is that the structure of the environment, the morphology of the species, the economics of species behavior, and the dynamics of population changes are the four essential ingredients of all interesting biogeographic patterns. Any good generalization will be likely to build in all these ingredients, and a bird pattern would only be expected to look like that of* Paramecium *if birds and* Paramecium *had the same morphology, economics, and dynamics, and found themselves in environments of the same structure.*

I am presenting here a nontraditional outline. My biases are evident also in other ways. I am devoted to the search for recurring patterns of plant and animal distributions and abundances, which I try to illustrate with the most appropriate examples I know. But these examples have a heavy bias to vertebrates and trees and to northeastern United States, Arizona, and Panama. They are not intrinsically better than could be drawn from invertebrates or from other places, and a reader with a different background is encouraged to supply his own material, and if he fails, to ask what is wrong with the principles proposed here, that they do not apply. Since a book committed to principles is doomed to early obsolescence while a book of pure observations is never out of date, the reader will find many places where the principles fall short of the desired generality and precision. In the spirit of science, he should discard a tentative generalization when he has a better one to suggest.

Part I
The Background

The full background needed for understanding the geographical patterns of ecology is too vast to put into a few chapters at the beginning of a book. Fortunately, however, most of the background is well known to anyone interested in biology or natural history, and here it is only necessary to include some topics that may be less familiar.

The first chapter sets the stage for discussing the geography of life by showing how the earth's rotations about its own axis and about the sun determine the large scale patterns of the world's climates. This is familiar ground to the geographer and, in part, to the physicist, but ecologists are less often exposed to it.

The next two chapters, on competition and predation and on optimal foraging strategies, cover topics treated in elementary courses. However, we shall need very advanced results and so offer background material that is far more thorough than appears in other books. The treatment is fully coherent without the mathematical sections presented in the Chapter 2 appendix but loses depth without them.

The final background chapter covers material belonging to evolution theory. Biologists in general know now how species split, but the detailed geographical stages are much less familiar. In Chapter 4, we give a brief account of the stages of speciation.

These four chapters provide the formal machinery for understanding the patterns described in Part II. The less formal knowledge and judgment automatically acquired by the field naturalist constitute the remaining equipment necessary for the proper understanding of these patterns.

Climates on a Rotating Earth

1

The distribution of the world's climates is too regular to be due to accidents of history. The world's deserts are mostly at 30°N or S latitude and usually on the west sides of continents. Just farther toward the poles on these west sides of continents we come upon "Mediterranean climates" and farther poleward yet, the heavy rainfall, especially in the winter, that causes the temperate rain forests of Washington and southern Chile. The equatorial regions are usually wet by frequent thunderstorms of short duration. These, and innumerable other patterns, suggest that we can understand the climate as a consequence of the earth's geography, particularly its rotation. The understanding of these relations is our goal in this first chapter.

Temperature

Our heat comes from the sun, and every point on the earth gets the same amount ($\frac{1}{2}$ year) of daylight each year (because, by symmetry, for each day longer than 12 hours in the summer there is a corresponding winter day that is shorter by the same amount). But the fact that two places have the same number of hours of daylight does not mean that they get the same amount of heat from the sun. Imagine a beam of light 1 inch in diameter, carrying energy, and striking a portion of the earth's surface that is perpendicular to the beam: the energy of the beam will be distributed over 1 inch of the earth. The same beam 1 inch in diameter striking at higher latitude will be spread over a greater area of the earth and so will give less heat per unit of area (see Figs. 1-1 and 1-2). Thus, as we all know, it not only averages hotter at the equator than at the poles but is also more uniformly hot, whereas the poles have a dark, very cold winter followed by a light, relatively warm summer. This requires no further elaboration.

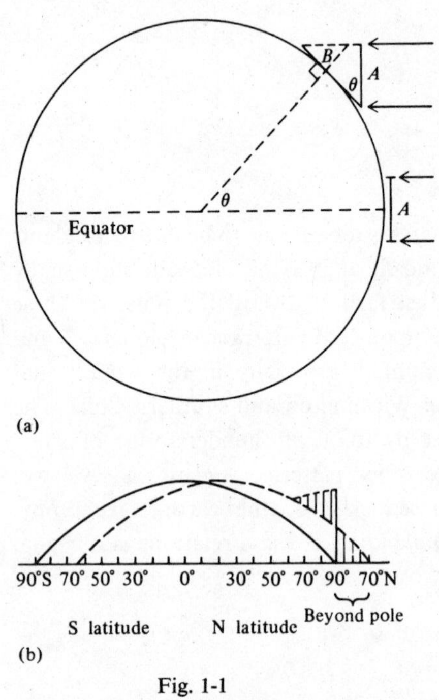

Fig. 1-1

(a) Two parallel beams of light as might come from the sun on the equinox (sun over equator). At latitude $\theta°$, the same amount of incident light falls on area $B = \frac{A}{\cos\theta}$. Since the intensity of light is the total radiation per unit area, it is proportional to the cosine of the latitude. (b) Intensity of light falling on the outer part of the earth's atmosphere at different latitudes. The solid line is at equinox, the dashed line at the summer solstice, June 21, when the sun is at 23°N latitude. Notice how the dashed curve folds back at the poles because the sun reaches 23° beyond the pole. The shaded areas are equal.

There is, however, an aspect of temperature that is not so obvious—the cooling with higher elevations. Why are tropical mountain tops as cool as temperate or even polar regions? Qualitatively the answer is quite simple. A parcel of air rising up a mountainside expands, like all its neighboring parcels, with the rarefied atmosphere. It does some work—uses some energy—in expanding, pushing the neighboring parcels aside and as all neighboring parcels are doing exactly the same, the energy cannot have come from them; it must have come from the parcel's own supply of energy, which is heat. In other words, the parcel must cool. The consequent rate of cooling, which we call the adiabatic lapse rate, is 5.5°F/1000 ft, or 10°C/km. "Adiabatic" means no external source of energy was involved. But not all lapse rates are *dry* adiabatic. If the air is condensing moisture as it cools, it gains some heat of condensation, which partially counteracts the cooling. Hence the "moist adiabatic" lapse rate (about 3°F/1000 ft, or 6°C/km) is less than the dry rate. Desert mountains cool faster with elevation than mountains in moist climates; and as the air on a mountainside cools enough to condense, the lapse rate changes from dry to moist. Figure 1-3 shows, for example, a mountain with air cooling at the dry rate as it begins ascending; at about 1 kilometer, when it is cool enough for moisture to condense, it changes to the wet rate. Often, on the lee side of a mountain there is no remaining moisture, so the descending air warms at the dry rate, leaving the valley air warmer on the lee than on the windward side. The derivation of the adiabatic lapse

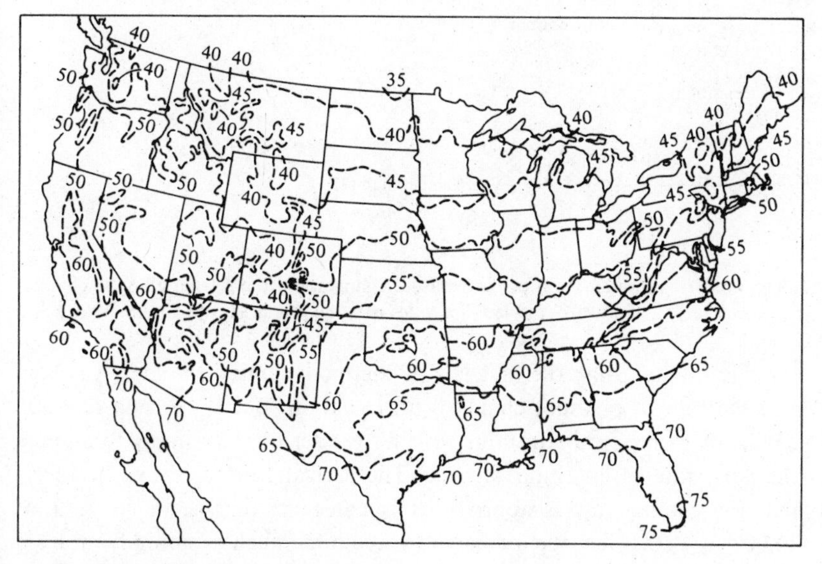

Fig. 1-2

Mean annual temperatures in degrees Fahrenheit in the United States (period, 1899–1938). From the 70° isotherm in Texas north to the 35° isotherm in North Dakota is about 1200 miles: just over 34 miles north per degree temperature rise. (From U.S. Department of Agriculture, 1941.)

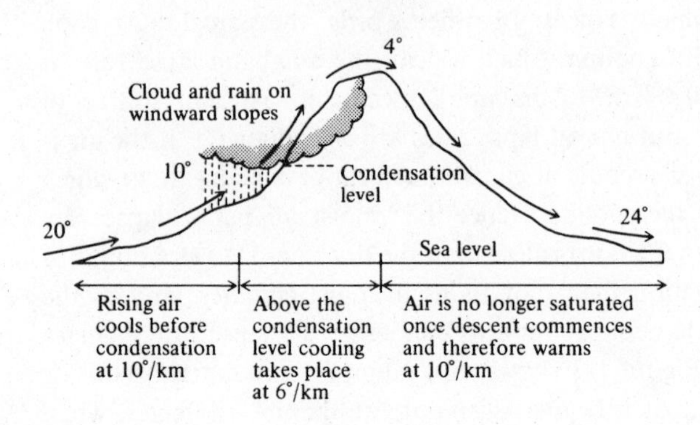

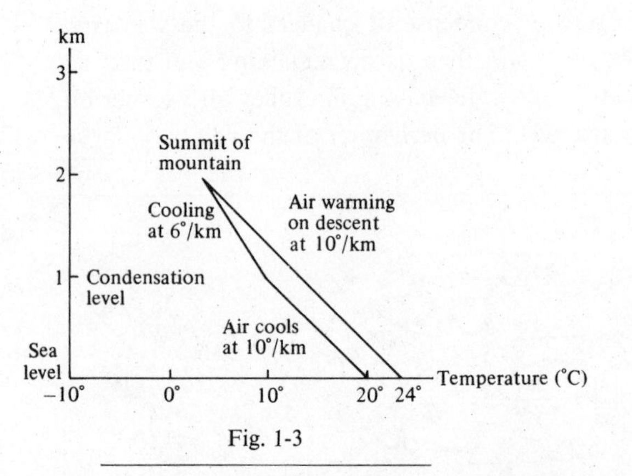

Fig. 1-3

Diagrammatic representation of temperature changes and the related climatic changes produced on the windward and leeward sides of a mountain. (From Flohn, 1969.)

rate* is simple, using the first law of thermodynamics, the ideal gas law, and the rate of pressure change with elevation. But since each of these has its own empirical constants, it is no greater burden simply to assume the lapse rate as an empirical rule. The derivation does show, however, that the cooling rate is approximately constant; that is, in going from 14,000 to 15,000 feet the air cools by about as much as going from 8000 to 9000 feet, and so on.

We now see why mountaintops are cooler than lower elevations. In fact, the 3°F cooling of 1000 feet of elevation on a moist

* See appendix to the chapter.

mountain is very roughly equal to the cooling of 100 miles of latitude, which has been called "Hopkin's bioclimatic law" (recall Fig. 1-2). But there is another lesson to be learned from the adiabatic lapse rate. We might say that under "normal" conditions the air cools at the adiabatic lapse rate as we go up in elevation. But we have all heard of "temperature inversions" and other abnormal temperature changes. Let us compare one of these with the normal adiabatic lapse rate, as we do in Fig. 1-4.

The basic picture is very simple: If heavy air is on top of light air, the heavy air will tend to sink through, hence heavy over light air is "unstable." Conversely, light air over heavy air is "stable," and light upper air is likely to stay in place.

We picture an atmosphere that has a nonadiabatic lapse rate for some reason, and first we consider one that cools faster than adiabatic with elevation. Now we suppose there is a small parcel of air being lifted by turbulence and cooling adiabatically of course as it rises. This air finds itself less cool than the neighboring air at each elevation and hence it is lighter and continues to rise, sucking up air behind it. Thus air cooling at a lapse rate faster than adiabatic is vulnerable to turbulent eddies of adiabatic air and is called "unstable." An atmosphere that cools more slowly than adiabatic with elevation is different. In this atmosphere a small eddy of rising air cooling adiabatically would be cooler and denser than surrounding air and hence would sink back down. Thus air which cools at a slower rate than adiabatic is "stable" and not vulnerable to eddies and other disturbances. Air that is warmer above than below is even more stable; this condition is called a temperature inversion and occurs when warm air over cold ocean or land cools at the zone of contact. Warm upper air over a cold ocean has no chance to descend and pick up its moisture. Warm air over a cold nocturnal desert is stable but gives way to unstable air, and "dust devils," as the desert warms in the daytime.

Atmospheric Global Circulation

When the sun's heat warms equatorial air, the air expands and rises. (On rising it cools, and since cold air holds less moisture, this causes the heavy tropical rains; here we anticipate our section on rainfall.)

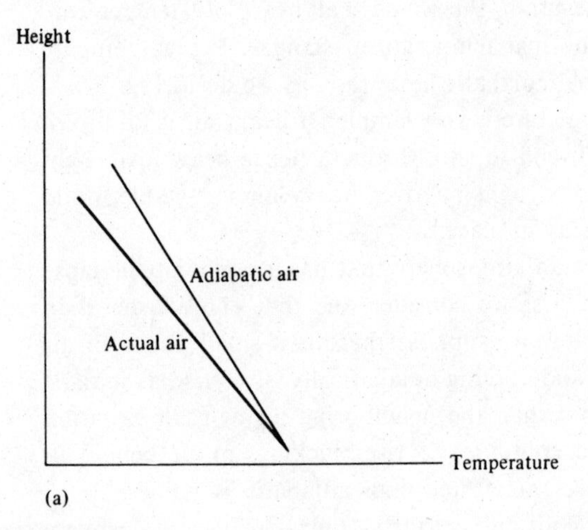

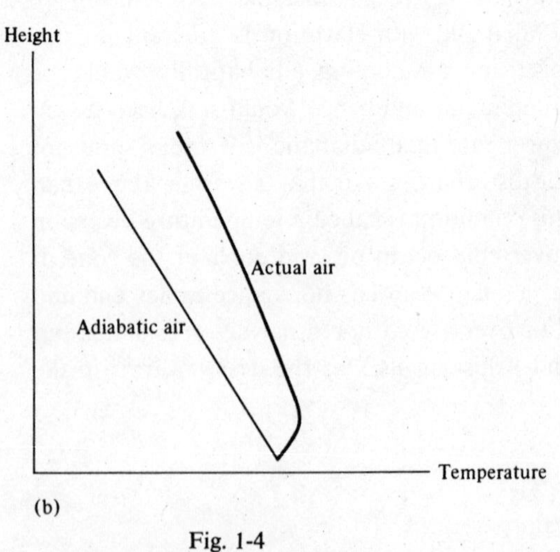

Fig. 1-4

(a) When the actual air cools more rapidly than adiabatic air, with height, it is unstable.
(b) When the actual air cools less rapidly than adiabatic it is stable. The figure shows a very stable temperature inversion, with warm air over colder.

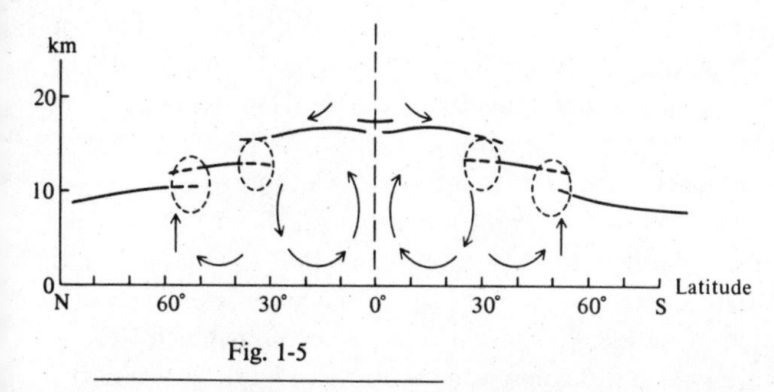

Fig. 1-5

Direction of circulation of the earth's atmosphere, shown by the arrows. The dashed circles are "jet streams." For a discussion see the text. (After Flohn, 1969.)

But the rising air must be replaced by other air, so we can infer surface winds on the earth rushing toward the equatorial regions to replace the rising air. (These are the trade winds, and we will see in the next section why they deflect to come from northeast and southeast.) The rising air does not go up forever, of course. Rather, it spreads out and, for no very clear reason, falls again at about 30°N and S latitude (see Fig. 1-5). There we have cool, dry air warming as it falls toward the earth; it can pick up more moisture and certainly has none to lose; hence we get our first hint of why deserts are usually at this latitude. The rest of the circulation pattern is hard to explain, but it at least has all the air going somewhere and coming from somewhere.

Coriolis Force

Hard as it is to account for the air movements of Fig. 1-5, they are the key to the explanation of the rest of the climate patterns. But we need one more tool, Coriolis force, which will show us why moving objects in the Northern Hemisphere deflect right and moving objects in the Southern Hemisphere deflect left.

The surface of the earth rotates from west to east, making the sun appear to rise in the east and set in the west. At the equator, the earth's surface is far from its axis of rotation, so it moves rapidly. Since the earth is about 24,000 miles around at the equator, the surface moves at a rate of 24,000 miles/day and so does any object on the surface. Farther north, at, say, 45°N latitude, the parallel of 45° latitude is only cos 45° = 0.707 times as long, which is just under 17,000 miles. Hence an object on the earth's surface at this latitude is moving at a velocity of 17,000 miles/day. In other words, at higher latitudes the surface moves more slowly.

Now picture an object moving north from the equator. While at the equator it had no east → west motion relative to the earth and so was, like the earth's surface, moving east at 24,000 miles/day. By the law of conservation of momentum it will tend to keep this much west → east momentum even when it reaches a more northern latitude, where the earth's surface itself is going more slowly. Hence, on moving north, our object will deflect right. Reversing the argument for another object, one moving from north latitudes toward the equator, we see it will tend to deflect west, but this can still be viewed as deflecting right. Therefore, we say that any object moving on the earth's surface in the Northern Hemisphere tends to deflect right; and similarly, moving objects in the Southern Hemisphere tend to deflect left.

Now we can explain winds and ocean currents. The trade winds rushing from the north toward the equator to replace the hot rising air deflect right and are thus coming from the northeast—they are the northeast trades. They pile up water as they blow along, so the Atlantic Ocean is deeper where the trade winds have pushed it into the Caribbean against Central America. In fact, the water on the Atlantic side of the isthmus of Panama averages a few feet higher than that on the Pacific side. This piled-up water escapes northward as the Gulf Stream. With its own Coriolis force it continually deflects right, around past England, down the west coast of Europe, and back toward the Caribbean. Wherever the trade winds can push the oceans against suitable barriers, they pile up the ocean water and cause ocean currents that move toward the poles on east sides of continents and then across the oceans and back toward the equator on the west sides of continents, the general deflection of the circulation always being that of Coriolis force (Fig. 1-6).

We shall return to these currents presently as they are of immense importance, but first we must discuss the cause of the prevailing westerly winds at temperate latitudes in both Northern and Southern hemispheres. We saw in Fig. 1-5 that there is in each hemisphere a poleward motion of the surface air from latitudes 30°N and S. Deflecting right in the Northern Hemisphere or left in the Southern, this motion will take a westerly component. Furthermore, the upper atmosphere is higher in midlatitudes than at high latitudes (refer to Fig. 1-5 again). In other words, above a given altitude there is more air at midlatitude than at high latitudes, and hence the pressure is greater at midlatitude. Air thus tends to move poleward in this zone and, deflecting right, forms a belt of westerly winds called the westerly jet. This belt of winds attains such a velocity that its Coriolis force southward balances the force northward that initiated the jet. Commercial planes often use these jet streams to get an extra 100 or 150 mph ground speed in their eastward flight. Hence both the surface and the upper air seem to have a strong westerly component north of 30°N latitude and south of 30°S latitude. (In the Southern Hemisphere at about 40°–50°S latitude there isn't much land to get in the way of these winds, and they are known as the "roaring forties." In 1854, the clipper ship *Champion of the Seas*, running east toward Australia in the roaring forties, went 465 nautical miles in 24 hours; but to beat westward against these winds especially through the Straits of Magellan proved too difficult for many a sailing ship.)

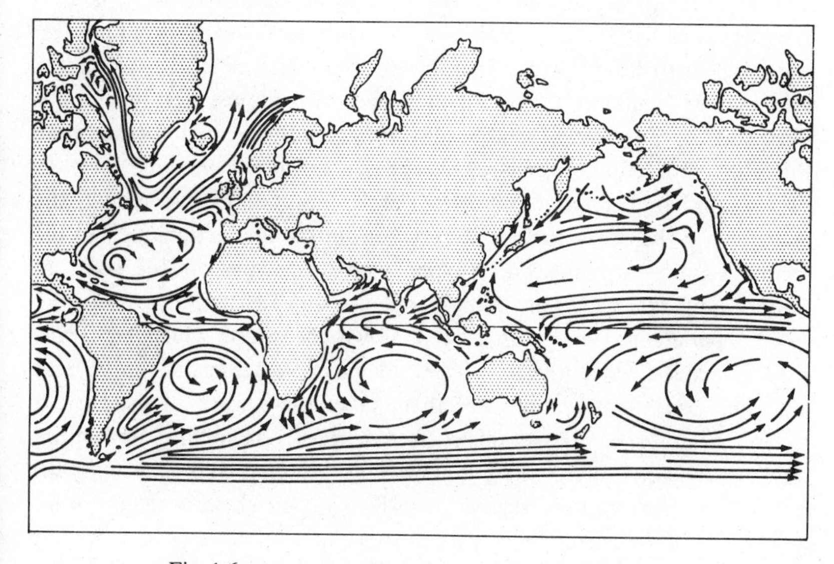

Fig. 1-6

Pattern of world's ocean currents. (From MacArthur and Connell, 1966.)

World Rainfall Patterns

Now we have the main elements in our explanation of the winds and the ocean currents. In combination they will account for the world's great rainfall patterns. In the trade wind zones the warm air picks up moisture from the oceans. When these moist trade winds hit a mountainous island or bit of mainland, their air goes up, cools, and loses large amounts of water. These same winds descending on the lee side of the mountains warm again and are thirsty for moisture; hence there is often a dramatic "rain shadow" on the southwest of tropical islands (in the Northern Hemisphere). Puerto Rico, for instance, gets heavy rain on its mountains but is a desert with succulent tree-like cactus species, reminiscent of Arizona, on its southwestern edge. Central America similarly tends to have a heavy rainfall where the trade winds hit the coast, but there is a rapid reduction in rainfall toward the Pacific slope, which is dry and even desert-like.

In the temperate zones our story involves the prevailing westerlies and the ocean current moving equatorward down the west sides of the continents. This current tends to be warmer than the land at high latitudes. Here the westerlies pick up moisture and, on cooling over the land, drop heavy rains. That is why the coast of British Columbia and that of southern Chile are drenched with almost continual rains (especially in the winter; then the offshore currents are much warmer than the land). Proceeding farther toward the equator along the same currents, two things change. First, the land gets warmer; second, the currents actually get colder at the ocean surface. The reason for the first is obvious, but the second requires explanation. Anyone who has swum off the coast of central California knows the water is cold, but why? There is, in this region, a wind paralleling the ocean current down the coast, which here runs from northwest to southeast. This wind moves faster than the current and thus exerts a dragging force that makes the surface of the current try to accelerate. Of course, this surface deflects right, pulling up bottom water to replace it. Thus there is upwelling, bringing cold (nutrient-rich) water from the bottom to the ocean surface. (The rightward deflection of the surface of the current causes an offshore motion of such velocity that *its* Coriolis force backward along the current just cancels the wind's force; hence a stronger wind causes stronger upwelling.)

Now our entire story changes. In discussing currents warmer than the land, we had westerlies picking up moisture from a warm ocean and dropping it on the cool land. But what happens if the ocean is cooler and the land warmer? When the land is warmer than the ocean, the westerlies do not drop rain but instead are thirsty for additional moisture. Only the cool mountains in such regions get rain. Northern California and Oregon are warmer than the ocean in the summer and get little rain; in the winter they are colder and have almost continuous rain. Southern California is almost always warmer than the ocean and, except in the mountains, gets virtually no rain. (San Diego gets only 10.11 in./year.) Baja California is even drier because it is warmer, and it is called desert although it is sparsely clothed with exotic succulents and cacti. (See the outstanding Sierra Club book *The Geography of Hope* for beautiful photographs and well-written commentary on Baja California.) Exactly the same pattern is repeated going north along the coast of Chile. Within that very long country one can proceed from cool rain forests so dense that Darwin had to walk over the top (without falling through), north through Mediterranean climate with citrus and grapes, like California, and farther north to the Atacama desert, the driest spot on earth. Some places there have never recorded rain. Of course, it is no accident that these deserts are on west sides of continents at about 30° latitude, and we can now summarize three main reasons for their location: (1) Cool dry air is descending and warming at 30° latitude (Fig. 1-5). This air will tend to pick up moisture if any is available; it won't lose any. (2) Prevailing westerlies blow in across a cold ocean. The temperature inversion, with cold air just above the cold ocean, is stable and the winds pick up little water. They then pass over warm land and would tend to pick up rather than lose moisture. (3) What little water the westerlies do have is usually lost as the winds pass over cool mountain ranges. The coastal ranges of California and Chile leave drier valleys farther inland, and the deserts of Arizona and California are in the rain shadow of the Sierras.

The rainfall as we proceed toward the eastern side of a continent is not just a consequence of the moisture the prevailing westerlies have picked up. We must also consider air masses. Looking at Fig. 1-5 again, we see that the air poleward of about 50° latitude is distinct from that equatorward. Between 30° latitude and 50° latitude the air moves poleward (and deflects toward the east) and is thus warm air. It is called tropical air even though it isn't really of tropical origin. In eastern United States and Canada, this warm air is often laden with moisture from the Gulf of Mexico or the Atlantic. This air mass meets the polar air mass that lies north of the westerly jet stream. If this were a complete picture, we would expect heavy rains only at latitude 50°. In fact, the jet stream meanders and breaks into eddies and vortices, leaving a belt of storms that spread over a much wider range of latitudes. The eastern parts of North America get most of their rains from cyclonic storms of this kind, distributed fairly evenly throughout the year.

Thunderstorms increase in importance toward the southeast, where parts of Florida get about 80 thunderstorms a year compared to 20–30 received by the northeast. When the ground heats during a hot day, it warms the lower air and causes an atmosphere that cools faster than adiabatically with elevation. Such an atmosphere, as we have seen, is unstable, allowing shafts of adiabatic air to penetrate and prosper. These set up strong updrafts, which are associated with the huge cumulonimbus

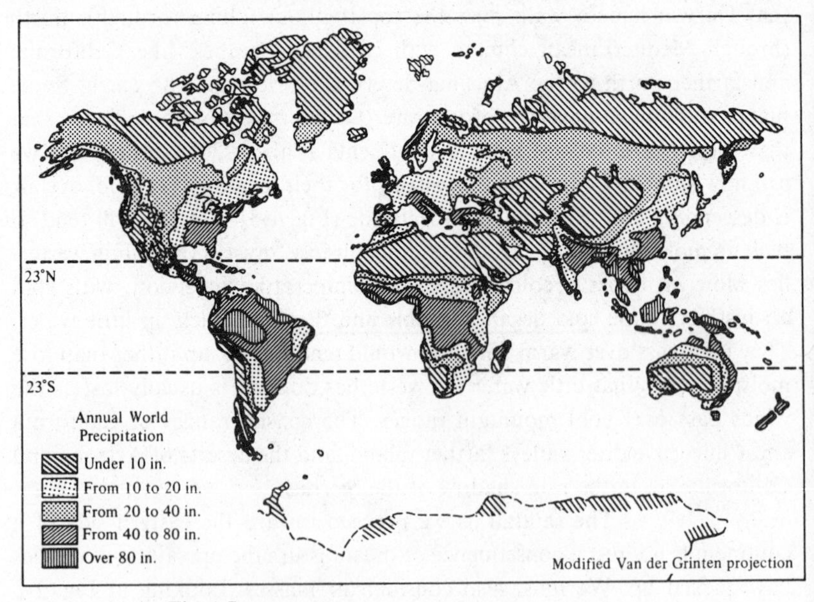

Fig. 1-7

Pattern of world precipitation. (From MacArthur and Connell, 1966, after Koeppe.)

thunderheads. At some point they are cooled enough to initiate a thunderstorm. Such storms may drop a total of 30–40 inches of rain per year on Florida and lesser amounts elsewhere. Even the Arizona deserts count on their summer rains from Gulf Coast maritime air persuaded to lose its moisture by thunderstorms in an unstable atmosphere. For the world pattern of precipitation, see Fig. 1-7.

Proximity to oceans or other large bodies of water that are heat reservoirs has a great moderating effect on temperature; the interiors of continents, far from oceans, are said to have a "continental" climate, having cold winters and hot summers. This is clear on comparing Wisconsin and Michigan. Wisconsin has very cold winters, but the westerlies blow on across Lake Michigan, which is a huge heat reservoir, and Michigan winters are much warmer.

There is at least one more important cause of rainfall, and it has produced the heaviest rains ever recorded. In summer, as continents warm faster than the oceans and the air above them rises, we get strong winds literally sucked in off the oceans. These are the monsoons. When they hit mountains, they lose prodigious amounts of water. Cherrapungi, India, received 366 inches (over 30 feet) of rain in one month, July 1861, as a result of exceptionally heavy monsoons.

One other point about rainfall should be mentioned. We who live in the temperate zone are familiar with "storms." We hear of one approaching from the west and whirling across the continent, and we prepare for several days of bad weather. Sometimes there may be as many as 10 days with rain every hour. Sometimes also these storms are accompanied by severe cold, freezing rain, or very strong winds. Table 1-1 shows a comparison of rainfall duration in temperate and tropics, indicating clearly that New Jersey storms are of longer duration than those on Barro Colorado Island, Panama Canal Zone. In the temperate regions and the very edge of the tropics there are so-called tropical storms that we call hurricanes and others call typhoons, and that often do severe damage on the east coast of the United States. Such storms as these—even the hurricanes—are largely unknown in the real tropics and never occur within 5° of the equator (Byers, 1954). There may be seasons of the year with and without trade winds; there may be wet seasons and dry seasons, and there may be daily rains. But except in regions of exception-

Table 1-1
Comparison of Rainfall Duration in Temperate and Tropics, 1967

Duration of rain (hr)	Number of rains: Trenton, N.J.	Number of rains: Barro Colorado Island, Panama Canal Zone
⩾1	186	415
⩾5	73	26
⩾10	38	2
⩾15	21	1
⩾20	11	1
⩾25	5	1
⩾30	2	0
Mean duration (hr)	6.242	1.945

ally heavy rainfall, these are usually of short duration, and even though much rain falls the sun is soon out again, the insects flying, and the birds feeding once more.

Appendix Derivation of Adiabatic Lapse Rate

Here we derive the dry adiabatic lapse rate from the gas law $pV = RT$ (p = pressure, V = volume, R is a constant, and T = absolute temperature) and from the first law of thermodynamics that says that when no heat is added to a volume of gas (i.e., adiabatic) $c\,dT = -p\,dV$ where c is the specific heat at constant volume and equals $\frac{5}{2}R$ and dT and dV are differentials of temperature and volume. We must also calculate how pressure changes with altitude by considering a column of gas of unit cross section. At height $h + dh$ the pressure will be lower than at height h by the weight of the cylinder of gas of unit cross section and height dh. This weight is $w\,dh$ where w is the weight of unit volume of gas at that altitude. $w = g\dfrac{M}{V}$ where g is the acceleration of gravity, 980.665 cm/sec^2, M is the molecular weight of the gas, and V is the volume occupied by 1 mole. In other words, $dp = -g\dfrac{M}{V}\,dh$ or $V\,dp = -gM\,dh$. Taking differentials of the gas law, $p\,dV + V\,dp = R\,dT$, whence the first law can be written $(c + R)\,dT = V\,dp$, and equating values of $V\,dp$, we finally get $(c + R)\,dT = -gM\,dh$, so the change of temperature T with respect to elevation h is given by

$$\frac{dT}{dh} = -\frac{gM}{c+R} = -\frac{980.665 \times 28.88}{\frac{5}{2}(8.214 \times 10^7) + 8.214 \times 10^7}$$

$$= -\frac{28321.605}{28.749 \times 10^7} = -985 \times 10^{-7}\,\frac{\text{deg}}{\text{cm}} = -9.85\,\frac{\text{deg}}{\text{km}}$$

The Machinery of Competition and Predation

2

Most people have a general idea what predation, parasitism, and microbial disease are: one species getting its nourishment from, and at the expense of, another. In this book we group all three together—in this wide sense—as "predation." An increase in the predator harms the prey, while an increase in the prey helps the predator.

The definition of competition is not as straightforward because the machinery is not as obvious. Some ecologists begin by listing various competitive mechanisms and giving a name to each. Here we use the alternative, wide definition of competition: two species are competing if an increase in either one harms the other. Any machinery that can have that effect will be called competition. For instance, species A and B can fight, or A can reduce B's food supply, or A can, by its own losses, increase B's predators. Provided that the effect is reciprocal, we will call all three competition.

In this chapter we present some of the effects competition and predation can have on species distributions and abundances. The proofs of our theoretical results appear in an appendix to the chapter. More specialized results of competition theory appear in appendixes to Chapters 7 and 8.

Both competition and predation appear now to be much more important in biogeography than people had formerly guessed. Notice that to be important they don't need to occur very often. For example, if food is scarce enough just 1 year in 20 to cause severe competition between two bird species, the inferior one is eliminated, and if the area of scarcity is large, the inferior species may take more than the next 19 years to reinvade. In other words, the better competitor may exclude the other species even though in a habitat where both could normally coexist an observer might only witness severe competition 1 year in 20. This is the main reason most evidence for competition is from biogeographers. The ecologist watching the populations may well not see them competing severely although the biogeographer has strong evidence that competition must sometimes occur.

We here review the experimental and theoretical nature of competition and predation, saving for later chapters the empirical cases where the biogeographer infers strong competition.

Experiments on Competition in Simple and Complex Environments

For years ecologists were plagued by an apparent paradox: many species seemed to coexist in nature but not in laboratory experiments. Gause (1934) put different species of microorganisms into media in bottles and only one species persisted. Park (1962) and others put different species of beetle or moth in flour in bottles and only one species persisted. Although the naturalist Grinnell (if he was aware of Gause's experiments) knew the explanation, it eluded experimentalists until some ingenious experiments of Crombie (1946). Crombie put *Tribolium confusum*

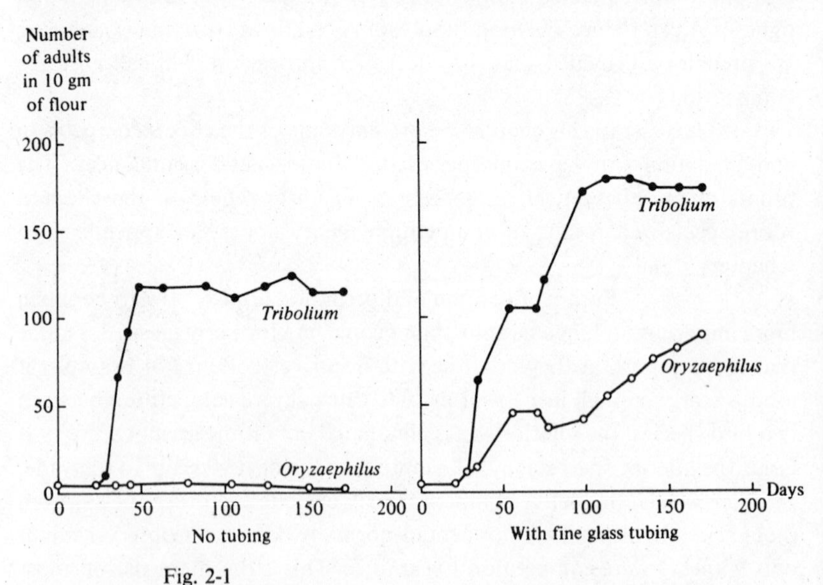

Fig. 2-1

Tribolium outcompetes *Oryzaephilus* in renewed flour unless fine glass tubing is added to the medium. The addition allows coexistence. (From MacArthur and Connell, 1966, after Crombie.)

and *Oryzaephilus suranimensis* together in flour and *Tribolium* won by competition. Then he put them in a mixed environment containing flour plus some broken wheat kernels and the species coexisted! Again he put the species together, this time in flour containing pieces of broken glass tubing. The tubing provided hiding places for *Oryzaephilus* and again he had two species coexist (Fig. 2-1). Grinnell had said that there must be two niches in the environment to have two species; hence he would have interpreted Crombie's experiments by pointing out that Crombie had added a second niche when he added the broken wheat kernels or the glass tubing.

There has arisen from these and similar experiments a widely accepted belief best called the principle of competitive exclusion, or Gause's principle (after G. F. Gause, who was one of the pioneer investigators of competition). The principle states that two species cannot coexist unless they are doing different things or, more baldly, unless there are two niches. Unfortunately, "niche" was never defined to make this a true and useful statement, so we must abandon that aspect of Gause's principle. However, people have recently tried to determine how different "different" must be for the species to coexist. In these terms the so-called competitive exclusion principle can be replaced by one that might say (its details have not yet been fully worked out), "Species that differ only in size seem to require that the larger be about twice as heavy as the smaller in order to coexist." This form is empirical (Hutchinson, 1959; Diamond, in press). Another form, that can be justified both empirically and theoretically, will be developed in a later section. Here we content ourselves with observing that the simple results of Gause's experiments misled ecologists for 40 years. Astonishingly few people, in spite of Crombie's lead, have tried to diversify their bottle experiments to hold more species. Instead, much effort has been spent trying to disprove the obsolete "competitive exclusion principle." The most recent such experiment was carried out by Ayala (1971), who carefully fitted linear bivariate expressions for the relative rates of increase of two very similar species of *Drosophila* and found that, although the fitted equations predicted that one species should outcompete the other, in fact the two coexisted in the laboratory. Although this of course proves that for such similar species of *Drosophila* these equations are not an adequate description of real events and that higher, nonlinear, terms would have to be added, it does not disprove anything important

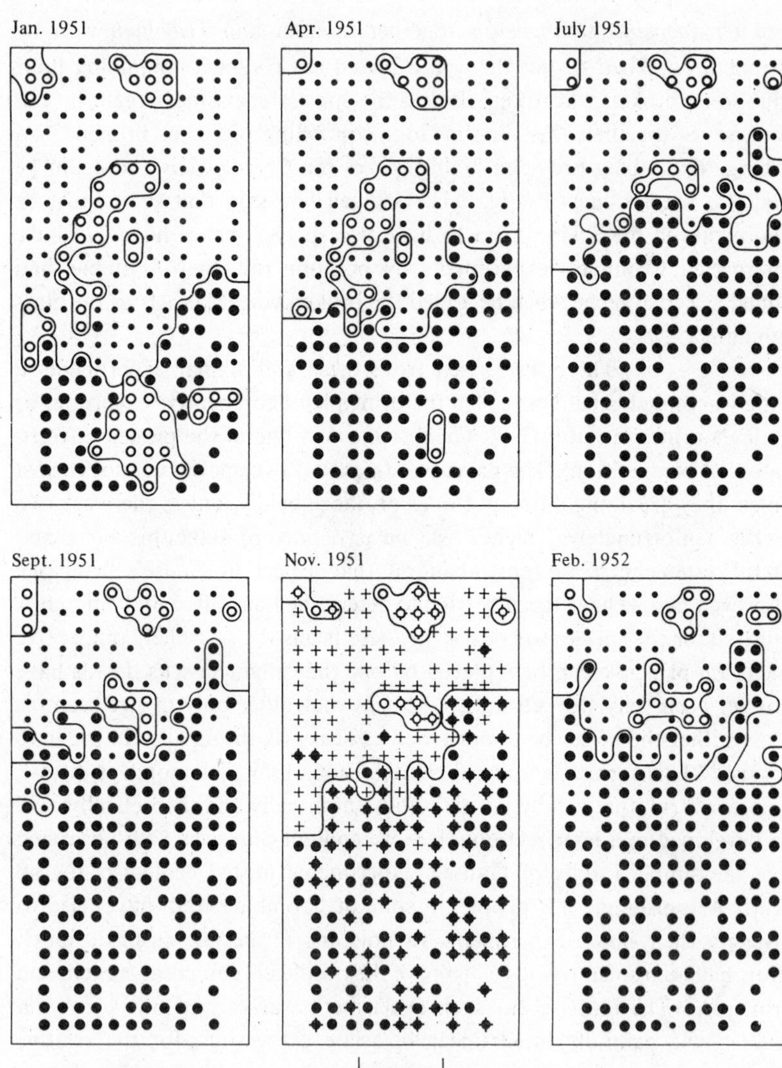
Jan. 1951
Apr. 1951
July 1951
Sept. 1951
Nov. 1951
Feb. 1952
50 yd
● Palm occupied by A. longipes
○ Palm occupied by O. longinoda
• Palm in which A. longipes and O. longinoda were absent
—Approx. boundary of territories occupied by A. longipes and O. longinoda
+ P. punctulata nesting at base of palm
◆ P. punctulata and A longipes present on palm
◇ P. punctulata and O. longinoda present on palm

about competition. Why haven't *Drosophila* geneticists, who are now becoming interested in ecology, tried putting very heterogeneous media in population cages to see if many species of *Drosophila* can be made to coexist? There is every reason to believe they would be successful. W. B. Heed (pers. comm.) and his colleagues have been rearing the *Drosophila* that emerge from rot pockets in cactuses. Different species of *Drosophila* seem to require different species of cactus, so that, for instance, *D. nigrospiracula* is found in three cactuses, including the saguaros; *D. mohavensis* are found in *Agria* and organ pipe (*Lemaireocereus thurberi*), *D. arizonensis* in *Rathbunia*, and *D. pachea* in senita (*Lophocereus schottii*). In fact, *D. pachea* needs a unique sterol produced in senita in order to develop and reproduce, and alkaloids present in senita are toxic to all other *Drosophila* species tested.

With our new perspective we look back at the early bottle experiments on competition as instructive because of their contrast with nature rather than their parallel to it. They proved that environmental heterogeneity is essential for coexistence by showing that in bottles where heterogeneity was absent, coexistence was rare if not impossible. But further experiments on coexistence with diverse media are essential if we wish to see how different the media must be to support two species.

Behavior and Competition

Wilson (1971) has documented a very large number of cases of competition with and without behavioral interactions (see Fig. 2-2). Following A. J. Nicholson and some other earlier authors, he says two species that compete by depleting the resource supply are having a "scramble," while two who compete by combat or interspecific territoriality are having a "contest." Although Wilson notes that the terminology is not well

Fig. 2-2

Exclusion of the ant *Oecophylla longinoda* by its competitor *Anoplolepis longipes* in a coconut plantation in Tanzania. The exclusion occurs through fighting at the colony level. In areas of sandy soil with sparse vegetation, *Anoplolepis* replaces *Oecophylla*, but where the vegetation is thicker and the soil less open and sandy, the reverse often occurs. A third species, *Pheidole punctulata*, is occasionally abundant but plays a minor role. (From Wilson, 1971, after Way.)

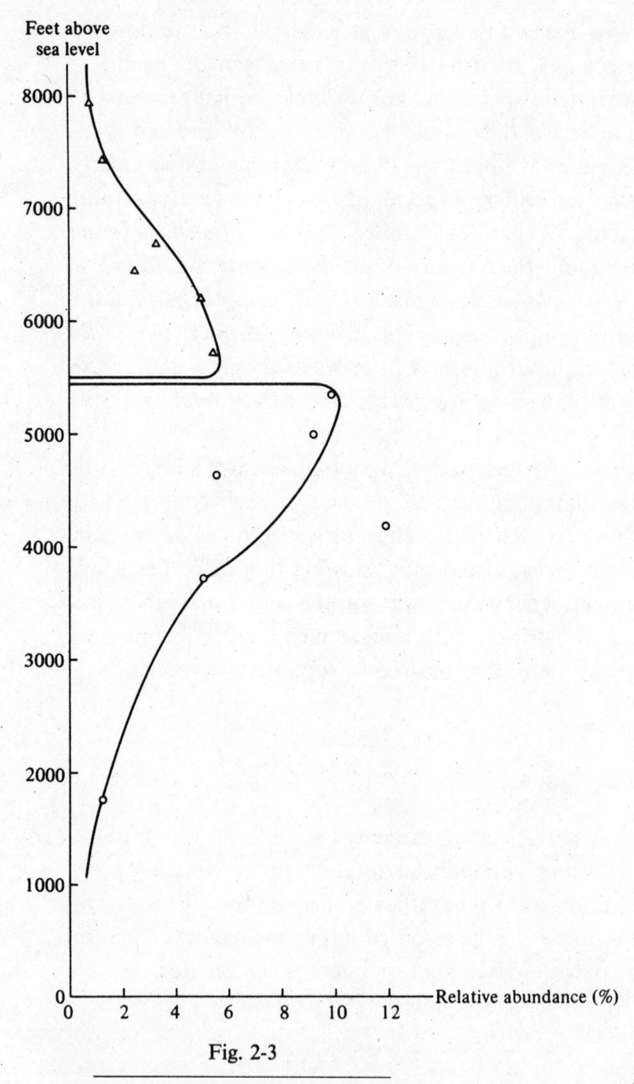

Fig. 2-3

Distribution of the warblers *Crateroscelis robusta* (△) and *C. murina* (○) on Mt. Karimui as a function of altitude. The figure gives the relative abundance in the whole avifauna—i.e., the percentage of bird individuals of all species estimated as being *C. robusta* or *C. murina*. Note that the two species replace each other sharply at 5400 ft and that each species reaches its maximum population density just above or just below this altitude. (After Diamond, in press.)

selected, there is a popular belief that the behavior competition is in some way more orderly, less chaotic, than competition by reducing resources: The behavior reduces a chaotic scramble to an orderly contest. Here we investigate this belief.

First we note that there is a natural priority: competition for resources can exist by itself, with no aggressive overtones, while aggressive competition would be pointless unless, in its absence, there would be resource competition; there would be no reward for the fighting if some new resource were not gained. Hence we conclude that behavior competition is a modification of an underlying resource competition. This is the principal reason why, contrary to the arguments of such popular writers as Robert Ardrey, aggressive behavior is far from universal in the animal kingdom.

Diamond (in press) has shown the abundances of two closely related bird species where their ranges abut on New Guinea mountains (Fig. 2-3). Presumably, the resources vary continuously on the mountainside and, under resource competition, these species would have overlapping ranges. By virtue of behavioral interactions (assumed to be interspecific territoriality) the two species have eliminated the zone of overlap completely. In what way is this an improvement? It may prevent mistakes in mate selection, and it substitutes for the overlap a more easily remembered arrangement of species, so that each species, by surrounding its members with others of their own kind, can be assured it is in the preferred habitat. Hopefully, each species is a better fighter in its preferred kind of habitat; but are we sure that the aggressive boundary coincides with the place where the more efficient forager shifts? It has been shown (MacArthur, 1970, and appendix to Chapter 8) that resource competition often leads to those abundances of the competing species that, in a certain sense made clear in the appendix, most fully utilize the production of resources. No alteration of these abundances according to which is locally the better fighter, then, can increase the resource utilization; rather it must *decrease* the effectiveness of the resource utilization, and the necessary fighting will decrease it even further.

We will now consider a probable sequence of events during the acquisition of aggressive competition by species that already have resource competition. Normally, without aggression, species A will be present by itself in some habitats, B will be present alone in other habitats, and both will be present in intermediate habitats. Now, what is the benefit of one species fighting the other? Species A should start a fight only in those places where it can improve its lot by so doing. Where will B fight back? Only where it will have some hope of winning. Like A, it will only take the offensive where it can improve its lot—and in so doing will reduce the overall foraging efficiency maximized by resource competition. But now we suppose A is a very good fighter and B is rather poor. Then it will benefit A to fight wherever it can make a living, and B will be reduced to the area where A is unable to persist. This would be a natural outcome of evolution, but it is only a "better" arrangement than the original resource competition if we define "better" to be "what evolution produces." The situation is reminiscent of the case in genetics in which segregation distorters automatically increase but confer no net benefit.

Species like ants with very highly developed social behavior would be expected to have wars. Wilson points out that better organization rather than sheer strength may determine the winner. For example, *Pheidole megacephala* worker ants calm down quickly after aggressive encounters and regroup at the feeding site by regaining and following their odor trails, while *Solenopsis globularia* workers recover more slowly and, as a result, lose control. Vertebrates seem much less sophisticated about their wars and usually determine the winner on the basis of a series of one-to-one conflicts.

The purpose of this section is not to belittle the importance of aggressive competition. Far from it: aggression seems to have been added to a vast number of competitive interactions, thereby substituting ability in combat for ability in utilizing resources as the determiner of the outcome. It is rather our purpose to point out that evolution will superimpose aggression onto resource competition if it benefits either species to fight. It need not benefit both. Thus for instance G. Orians (pers. comm.) has found that where yellow-headed blackbirds (*Xanthocephalus xanthocephalus*) are present they drive the smaller red-winged blackbirds (*Agelaius phoeniceus*) out of parts of the marshes. Fortunately for the redwings, the yellow-headed blackbirds need a more productive marsh, or parts of a marsh, leaving the less productive areas in the sole possession of the smaller species. The redwings seem to derive no benefit from this combat.

Results of Theory of Competition

We now summarize the results, proved in two alternate treatments in the chapter appendix, suggested by the mathematical theory of competition. The competition equations require a good understanding of calculus and probability and for this reason are presented separately.

Since the premises for the theory are not universally valid, it is best to regard the results as hypotheses to be tested rather than as established principles. The results suggest observations to make in the field and thus to provide evidence for the importance of competition.

Competitor similarity. The more similar the competing species, the smaller their zone of geographic or habitat overlap and the more vulnerable their coexistence in this zone of overlap to environmental or other variations. The more similar the noncompetitors, the larger their overlap.

Diffuse competition. Several competitors can much more easily outcompete and eliminate a species than a single competitor can. This is called diffuse competition. Even three species may be packed so tightly that the middle one has almost no chance.

Resource overlap. The coexistence of two or more competitors becomes rapidly more precarious as the distance between their resource mean values approaches $\sqrt{2}$ times their standard deviation (see Fig. 2-7, p. 44).

Geographic sequences of competitors. Along a geographic continuum two competitors may be found in either of the following sequences: (1) one species, then both, then the other; (2) one species, then a vacant zone, then the other. Situation (2) is more likely to occur in an unproductive environment. With behavioral or aggressive competition added, two species may have sharply abutting ranges.

For further results and mathematical treatments, see pages 119–120; Chapter 5 appendix; Chapter 8, Appendix 1; pages 247–251.

Experiments on Predation in Simple and Complex Environments

The history of experiments on predation parallels that of competition. Predators and their prey seem to have no trouble coexisting in nature but great trouble coexisting in the laboratory. Gause (1934) put *Paramecium caudatum* and their predators *Didinium nasutum* in bottles together and either the *Didinium* ate all the *Paramecium* and then starved, or a few *Paramecium* found hiding places and reemerged to increase after the *Didinium* had all starved.

Notice that predator-prey interactions have a built-in time lag. When the prey are commonest, the predators are increasing the fastest, but there is some delay before they reach their commonest, and by that time they have overeaten the prey, which have decreased. The predators continue increasing after the prey have reached their maximum and continue decreasing after the prey have reached their minimum. This last often proves fatal in the laboratory, with the already small predator population coasting down to zero. But there is an even more important reason for the failure of coexistence in the laboratory. The environment is so simple and packed with food that a predator can find every prey item. It is simply a question of seeing them all and pursuing them. There are few natural environments of this sort. Paine (1966) has shown that starfish can eat every mollusc and barnacle in a local area of intertidal zone. Within an enclosure in the intertidal preventing emigration and immigration the starfish would probably exterminate their prey as laboratory predators do. But what distinguishes the whole intertidal zone from a laboratory bottle is its size and the lack of synchrony of its parts. Whenever one patch is swept clean by starfish, another, swept clean a year earlier and with all its starfish emigrated, is making a recovery. Most natural environments are not so easily searched. A bird searching the forest foliage for scattered, hidden insects has no chance of finding them all. In an attempt to imitate the complexities of a real environment in the laboratory, Huffaker (1958) raised predator mites and prey mites in a complex environment of oranges (the food of the prey mites) with deliberately added baffles to motion and other complexities such as bridges connecting the oranges. The more complex he made the environment, the longer predator and prey persisted together.

There is no point in discussing here the dynamics of predator-prey interactions, since population changes in time are not explicitly part of biogeography. (For the dynamics, see Rosenzweig and MacArthur (1963) and Maly (1969).) We will note only the fairly self-evident conditions that enable prey to keep from being exterminated. An efficient predator can exterminate prey, as we have seen. There are two principal kinds of environment in which this is prevented.

1. If the prey have a refuge in which the predator cannot hunt, the prey may survive. Of course, it must be a refuge sufficiently large so that all the life history of the prey can be carried out in it. If it is only a refuge for spending the night, say, it will decrease the effectiveness of the predator, which helps some, but it will not guarantee the survival of the prey.

2. If the environment, or the predators themselves, imposes an upper limit to the number of predators, the prey may be saved. The requirement, of course, is that the upper level of the predators not be high enough to eliminate the prey.

Results of Theory of Predation

Here, as for the treatment of competition, we present a summary of the results that can be derived from a mathematical consideration of predation. The proofs are presented in the chapter appendix.

Predator-prey oscillations. Predator-prey interactions tend to drive oscillations in both predator and prey populations. These oscillations tend to damp out if the predators are relatively inefficient or if the prey are short of food. The oscillations are prevented from causing extinctions if either the predators have a low ceiling on their populations or the prey have a satisfactory predator-free hiding place.

Overexploitation. Self-renewing resources can be overexploited to the detriment of the predator's population, and this overexploitation will be a natural consequence of competition among the predator species. Certain other kinds of resources, such as fallen fruit, cannot be overexploited in the short run (although in the long run new fruit-bearing trees might be prevented from replacing the old ones).

Effects of switching resources. A predator who switches to the commonest suitable resource can keep all resources rare. Thus if there are few resource species and predators keep them all rare, there may be room for more resource species.

Appendix
Theory of Competition and Predation

Curiously enough, the basic theory of competition was worked out by Volterra, the mathematician, before any of the experiments were carried out. After a distinguished career in mathematics, Volterra was interested in these problems by a son-in-law investigating fisheries biology. The theory directly inspired the experiments of Gause, and since that fulfilled one of the two principal goals of theory, one cannot consider Volterra's theory a failure, despite some limitations. The other main goal of theory is to inform us when we have a coherent explanation with all necessary ingredients. Volterra's theory did this too. It showed that simple equations might lead to coexistence or to one species replacing another, and it told us which to expect. In these respects, the Volterra theory that we are going to describe and extend has been an unqualified success. However, it does have its limitations: It is probably never literally correct. Very likely no population ever grows exactly according to Volterra's equations. Ecologists who use them are following reasoning something like this: The true, correct equations are probably "near to" the Volterra equations, and the behavior of such equations will be "near to" the behavior of the solutions of Volterra equations. Nearby equations have nearby solutions. They further reason that the stability and coexistence demonstrated for Volterra equations will not be altered by small changes. (The mathematical version of these beliefs is the statement that the equations are "structurally stable" in the sense of Thom (1970).) The really good mathematician might be able to find necessary and sufficient conditions for coexistence of many species without beginning with a specific system like Volterra's, but the ecologist and biogeographer find it easier to use specific equations, remembering that they are only "near to" the truth. And, of course, there is no guarantee they are always even "near to" the truth. In many cases they doubtless are; in other situations they may be hopelessly far from the truth and should not be used. The ecologist's judgment based upon understanding of the nature of the equations and upon field experience with the organisms being described is essential in assessing whether the equations are suitable.

Mathematical Treatment of Competition Theory

The competition equations and conditions for two-species coexistence. Let X_1 be the population of species 1 and X_2 the population of species 2. Then $\frac{1}{X_1}\frac{dX_1}{dt}$ is the per capita rate of increase of species 1 and $\frac{1}{X_2}\frac{dX_2}{dt}$ is the per capita, because of the $\frac{1}{X}$, rate of increase of the second species. (These are equivalent to interest rates in a bank; if you have N dollars in a bank, then $\frac{1}{N}\frac{dN}{dt}$ is the interest rate, expressed as a decimal, that the bank provides.) Volterra assumed these rates of increase were determined by the species' own abundance as well as by the competitor's abundance.

$$\frac{1}{X_1}\frac{dX_1}{dt} = f_1(X_1, X_2) \qquad \frac{1}{X_2}\frac{dX_2}{dt} = f_2(X_1, X_2)$$

In fact Volterra assumed the functions f_1 and f_2 were linear in X_1 and X_2:

$$\frac{1}{X_1}\frac{dX_1}{dt} = \frac{r_1}{K_1}[K_1 - X_1 - \alpha X_2] \qquad \frac{1}{X_2}\frac{dX_2}{dt} = \frac{r_2}{K_2}[K_2 - \alpha' X_1 - X_2] \tag{1}$$

These are written in the usual way to give the constants a readily interpreted meaning. First, when X_1 and X_2 are very small, $\frac{1}{X_1}\frac{dX_1}{dt} = r_1$ and $\frac{1}{X_2}\frac{dX_2}{dt} = r_2$, so r_1 and r_2 are called "intrinsic rates of natural increase," the per capita rates of increase of unimpeded growth. When $X_2 = 0$ but X_1 is allowed to grow until $X_1 = K_1$, then $\frac{1}{X_1}\frac{dX_1}{dt} = 0$; hence K_1 is the asymptote for X_1 when X_2 is absent. Similarly, K_2 is the asymptote of X_2 when X_1 is absent. These K's are called "carrying capacities" of the environment for the species. Finally, we see that an increase in X_1 reduces $\frac{1}{X_1}\frac{dX_1}{dt}$, and that an increase in X_2 reduces $\frac{1}{X_1}\frac{dX_1}{dt}$ by exactly α times as much (α is usually less than 1). Alpha is called the competition coefficient of X_2 on X_1. Similarly α' is the competition coefficient of X_1 on X_2. When $\alpha = 0$, X_2 does not affect X_1's growth, while when $\alpha = 1$, an individual of species 2 is exactly equivalent to an individual of species 1 in its effect on $\frac{1}{X_1}\frac{dX_1}{dt}$.

Now, although Eqs. (1) are known to have a solution, they have never been solved; that is, there are no known functions $X_1(t)$, $X_2(t)$ that can be written explicitly and that satisfy Eqs. (1). However, we can learn all we need to know about the solutions even if we can't write them down. We would like to know whether the two species can come to coexist or one outcompetes the other. Simple biology comes to our rescue here: X_1 persists if and only if it can invade a community of X_2 alone at equilibrium, and X_2 persists if and only if it can invade a community containing X_1 alone at equilibrium. When X_2 is alone, it reaches K_2 at equilibrium, so X_1 can invade if and only if $\frac{1}{X_1}\frac{dX_1}{dt} > 0$ when X_1 is rare (i.e., near zero) and $X_2 = K_2$. From Eqs. (1), this means X_1 can invade if and only if $K_1 - \alpha K_2 > 0$ or $\frac{K_1}{K_2} > \alpha$. Similarly, X_2 can invade an equilibrium community of X_1 alone if $\frac{1}{X_2}\frac{dX_2}{dt} > 0$ when X_2 is very near zero and $X_1 = K_1$; i.e., if and only if $K_2 > \alpha' K_1$ or $\frac{1}{\alpha'} > \frac{K_1}{K_2}$. Combining these two results, each species can invade the other and so both persist if and only if

$$\frac{1}{\alpha'} > \frac{K_1}{K_2} > \alpha \tag{2}$$

There is another way to get the same result in a few lines. We shall have to use the method on page 54 so we introduce it here. Equations (1) are at equilibrium if the terms in brackets are zero.

$$[K_1 - X_1 - \alpha X_2] = 0 \qquad K_1 = X_1 + \alpha X_2$$
$$\text{or}$$
$$[K_2 - \alpha' X_1 - X_2] = 0 \qquad K_2 = \alpha' X_1 + X_2$$

These can be written in column vector form:

$$\begin{pmatrix} K_1 \\ K_2 \end{pmatrix} = \begin{pmatrix} 1 \\ \alpha' \end{pmatrix} X_1 + \begin{pmatrix} \alpha \\ 1 \end{pmatrix} X_2$$

This has a solution in which both X_1 and X_2 are positive if and only if $\begin{pmatrix} K_1 \\ K_2 \end{pmatrix}$ can be written as a *positive* linear combination of $\begin{pmatrix} 1 \\ \alpha' \end{pmatrix}$ and $\begin{pmatrix} \alpha \\ 1 \end{pmatrix}$. This means the vector $\begin{pmatrix} K_1 \\ K_2 \end{pmatrix}$ lies between the vectors $\begin{pmatrix} 1 \\ \alpha' \end{pmatrix}$ and $\begin{pmatrix} \alpha \\ 1 \end{pmatrix}$. (When it does not lie between them, it has to be written as $\begin{pmatrix} 1 \\ \alpha' \end{pmatrix} m + \begin{pmatrix} \alpha \\ 1 \end{pmatrix} n$ where either m or n is negative.) So we can say both X_1 and X_2 are present (i.e., positive) at equilibrium if and only if the vector $\begin{pmatrix} K_1 \\ K_2 \end{pmatrix}$ lies between $\begin{pmatrix} 1 \\ \alpha' \end{pmatrix}$ and $\begin{pmatrix} \alpha \\ 1 \end{pmatrix}$. (This form has immediate application to many simultaneously competing species and would take the form in which the K column lies in the convex cone generated by the α column vectors.)

The main conclusion to draw is that when α and α' are near 1, the ratio of K's must be very precise to allow coexistence. If, for instance, $\alpha = \alpha' = 0.9$, then we must have $1.1 = \frac{1}{0.9} > \frac{K_1}{K_2} > 0.9$, while if $\alpha = \alpha' = 0.2$, $\frac{K_1}{K_2}$ can vary between the much wider limits $5 = \frac{1}{0.2}$ and 0.2. Now, very similar species have α values near to 1 by the very definition of α; if they are truly similar, each must depress the other's growth almost as much as its own. Hence we can reframe our conclusion: Very similar species coexist only under very precise $\frac{K_1}{K_2}$ ratio, while dissimilar species coexist more easily. This is the kind of conclusion that can be drawn directly from competition equations. It does tell us how similar species can be and still coexist but says it in terms of K's and α's. Unless we know how to measure these, we cannot test or use the results. The reason we have no explicit instructions about calculating K and α is that we have said nothing about the machinery of competition. If we postulate just how the competition is acting, we will get a recipe for calculating K and α. We do this next.

The calculation of K and α. Let us suppose the competition is really acting by X_1 and X_2 eating the same foods. Whichever eats a bit first deprives the other. To be fully explicit, so as to get a fully explicit formula for K and α, let us suppose the consumers, X_1 and X_2, are limited by resource supply alone and grow according to equations of the form

$$\frac{1}{X_1}\frac{dX_1}{dt} = C_1\,[a_{11}w_1R_1 + a_{12}w_2R_2 - T_1]$$
$$\frac{1}{X_2}\frac{dX_2}{dt} = C_2\,[a_{21}w_1R_1 + a_{22}w_2R_2 - T_2] \qquad (3)$$

Here R_1 and R_2 are the quantities of two resources, w_1 and w_2 are their weights per unit of quantity, a_{ij} is the probability that during a unit of time a given individual of species i encounters and eats a given individual of resource j, T_1 and T_2 are the weights of food needed for maintenance without population growth, and C_1 and C_2 are the factors governing conversion of grams of food into new individuals. Thus the per capita rate of increase is proportional to the excess of resource eaten over resource needed for maintenance.

We must also assume some form of resource renewal; if resources renewed instantly, species 1, by eating a resource, would not deprive species 2 so there could be no competition. Hence, it is clear that the form and rate of resource renewal will be critical. Suppose, to be explicit, that in the absence of consumers each resource grows "logistically"; i.e., according to $\frac{1}{R}\frac{dR}{dt} = \frac{r}{K}[K - R]$. Then in the presence of consumers we have

$$\frac{1}{R_1}\frac{dR_1}{dt} = \frac{r_1}{K_1}[K_1 - R_1] - a_{11}X_1 - a_{21}X_2$$
$$= \frac{r_1}{K_1}\left[K_1 - R_1 - \frac{K_1}{r_1}a_{11}X_1 - \frac{K_1}{r_1}a_{21}X_2\right]$$
$$\frac{1}{R_2}\frac{dR_2}{dt} = \frac{r_2}{K_2}[K_2 - R_2] - a_{12}X_1 - a_{22}X_2 \qquad (4)$$
$$= \frac{r_2}{K_2}\left[K_2 - R_2 - \frac{K_2}{r_2}a_{12}X_1 - \frac{K_2}{r_2}a_{22}X_2\right]$$

(It must be emphasized that had we described resources with a different pattern of renewal, we would conclude with different results. If, for instance, R_1 is a kind of fruit falling from old trees, then, in the absence of predation, $\frac{dR_1}{dt}$ = constant, say D, and so with predation

$$\frac{1}{R_1}\frac{dR_1}{dt} = \left[\frac{D}{R_1} - a_{11}X_1 - a_{21}X_2\right] \tag{4'}$$

This Eq. (4′) would then be substituted for (4) in what follows.)

We are interested in equilibrium values of X_1 and X_2, and R_1 and R_2, so we set all brackets equal to zero in Eqs. (3) and (4). From Eqs. (4), at equilibrium,

$$\begin{aligned} R_1 &= K_1 - \frac{K_1}{r_1}a_{11}X_1 - \frac{K_1}{r_1}a_{21}X_2 \\ R_2 &= K_2 - \frac{K_2}{r_2}a_{12}X_1 - \frac{K_2}{r_2}a_{22}X_2 \end{aligned} \tag{5}$$

Now we set the brackets of Eqs. (3) equal to zero and substitute the values of R_1 and R_2 from Eqs. (5):

$$a_{11}w_1\left(K_1 - \frac{K_1}{r_1}a_{11}X_1 - \frac{K_1}{r_1}a_{21}X_2\right) + a_{12}w_2\left(K_2 - \frac{K_2}{r_2}a_{12}X_1 - \frac{K_2}{r_2}a_{22}X_2\right) - T_1 = 0$$

$$a_{21}w_1\left(K_1 - \frac{K_1}{r_1}a_{11}X_1 - \frac{K_1}{r_1}a_{21}X_2\right) + a_{22}w_2\left(K_2 - \frac{K_2}{r_2}a_{12}X_1 - \frac{K_2}{r_2}a_{22}X_2\right) - T_2 = 0$$

Collecting constant terms, terms in X_1, and terms in X_2, this becomes

$$0 = [a_{11}w_1K_1 + a_{12}w_2K_2 - T_1] - \left\{\frac{K_1}{r_1}a_{11}{}^2w_1 + \frac{K_2}{r_2}a_{12}{}^2w_2\right\}X_1 - \left\{\frac{K_1}{r_1}a_{11}a_{21}w_1 + \frac{K_2}{r_2}a_{12}a_{22}w_2\right\}X_2$$

$$0 = [a_{21}w_1K_1 + a_{22}w_2K_2 - T_2] - \left\{\frac{K_1}{r_1}a_{21}a_{11}w_1 + \frac{K_2}{r_2}a_{22}a_{12}w_2\right\}X_1 - \left\{\frac{K_1}{r_1}a_{21}{}^2w_1 + \frac{K_2}{r_2}a_{22}{}^2w_2\right\}X_2$$

These look almost like the brackets of Eqs. (1), set equal to zero; to get them into exactly that form, we divide the first by the coefficient of X_1 and the second by the coefficient of X_2:

$$0 = \left(\frac{a_{11}w_1K_1 + a_{12}w_2K_2 - T_1}{\frac{K_1}{r_1}a_{11}{}^2w_1 + \frac{K_2}{r_2}a_{12}{}^2w_2}\right) - X_1 - \left\{\frac{\frac{K_1}{r_1}a_{11}a_{21}w_1 + \frac{K_2}{r_2}a_{12}a_{22}w_2}{\frac{K_1}{r_1}a_{11}{}^2w_1 + \frac{K_2}{r_2}a_{12}{}^2w_2}\right\}X_2$$

$$0 = \frac{a_{21}w_1K_1 + a_{22}w_2K_2 - T_2}{\frac{K_1}{r_1}a_{21}{}^2w_1 + \frac{K_2}{r_2}a_{22}{}^2w_2} - \left\{\frac{\frac{K_1}{r_1}a_{21}a_{11}w_1 + \frac{K_2}{r_2}a_{22}a_{12}w_2}{\frac{K_1}{r_1}a_{21}{}^2w_1 + \frac{K_2}{r_2}a_{22}{}^2w_2}\right\}X_1 - X_2 \tag{6}$$

Equations (6) are now the terms in brackets of Eq. (1), set equal to zero, so we have achieved our first goal; we know explicitly how to calculate the K's and α's of Eq. (1) [see Ed. note on p. 58]:

$$K_1 = \frac{\sum_j a_{1j}w_jK_j - T_1}{\sum_j \frac{K_j}{r_j}w_ja_{1j}{}^2} \qquad K_2 = \frac{\sum_j a_{2j}w_jK_j - T_2}{\sum_j \frac{K_j}{r_j}w_ja_{2j}{}^2}$$

$$\alpha = \frac{\sum_j \frac{K_j}{r_j}w_ja_{1j}a_{2j}}{\sum_j \frac{K_j}{r_j}w_ja_{1j}{}^2} \qquad \alpha' = \frac{\sum_j \frac{K_j}{r_j}w_ja_{1j}a_{2j}}{\sum_j \frac{K_j}{r_j}w_ja_{2j}{}^2} \tag{7}$$

Equations (7) are taken directly from comparing Eqs. (6) with (1) and are written in summation notation so that they can be used for any numbers of consumers or resources. Notice particularly that the numerators of α and α' are identical; if the denominators are also equal, then $\alpha = \alpha'$. If Eqs. (4) had been replaced by ones like (4′) and these solved for R_1 and R_2 to be substituted into (3), we would have gotten different K's of course, and the equation would no longer have been linear in the X's. However, the coefficient of the first term in the Taylor series—as an approximation to the α components—would have been identical in form to the α of Eqs. (7). Instead of $\frac{K_j w_j}{r_j}$ as a weighting term, there would have been another.

Relation of α to the subdivision of resources among species. We now have, for a particular pattern of resource renewal, explicit recipes for the K's and α's of inequality (2). The α's are of particular importance and we here examine their form. We picture the resources arranged along a line j (say, from smallest to largest or from lowest to highest) and consider two species, 1 and 2, that differ in this way: 1 eats smaller food and 2 eats larger, with some overlap; or 1 eats from lower in the trees and 2 eats from higher, with some overlap. Let us then plot the degree of utilization of each resource by each species,

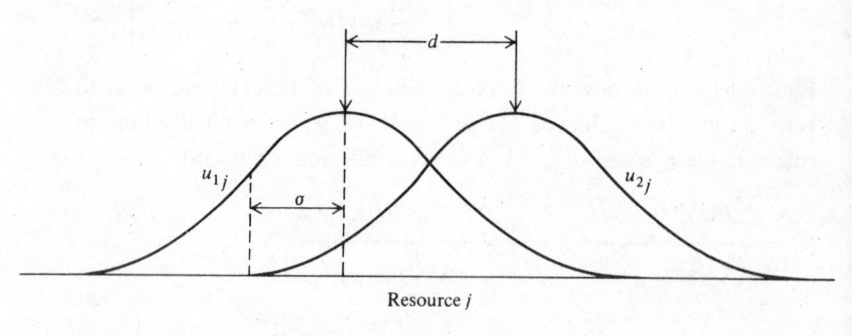

Fig. 2-4

Utilizations u_{1j} and u_{2j} of two species plotted along a resource coordinate. d is the distance between the means of the u curves, and σ is the standard deviation of either. If, for instance, j is resource size, the two means are the average food size of the consumer species and σ is the standard deviation of the food size of the consumers.

defined as $u_{1j} = a_{1j}\sqrt{\frac{K_j w_j}{r_j}}$ and $u_{2j} = a_{2j}\sqrt{\frac{K_j w_j}{r_j}}$, against this resource coordinate j as in Fig. 2-4. The u's are just abbreviations for the longer expressions on the right. We notice that $\frac{\sum_j u_{1j}u_{2j}}{\sum_j u_{1j}^2} = \alpha$ and we would like to relate this to the distance d in the figure. In the special but probably common case of the normal distribution where $u_{1j} = Ce^{-j^2/2\sigma^2}$ and $u_{2j} = Ce^{-(j-d)^2/2\sigma^2}$, $\alpha(d)$ is particularly simple to derive:

$$\alpha(d) = \frac{C^2 \int_{-\infty}^{\infty} e^{-j^2/2\sigma^2} \cdot e^{-(j-d)^2/2\sigma^2}\, dj}{C^2 \int_{-\infty}^{\infty} [e^{-j^2/2\sigma^2}]^2\, dj}.$$

Now $e^{-j^2/2\sigma^2} \cdot e^{-(j-d)^2/2\sigma^2} = e^{-d^2/4\sigma^2} \cdot e^{-(j-d/2)^2/\sigma^2}$, and $[e^{-j^2/2\sigma^2}]^2 =$

e^{-j^2/σ^2}, whence $\alpha(d) = \dfrac{e^{-d^2/4\sigma^2} \int_{-\infty}^{\infty} e^{-(j-d/2)^2/\sigma^2}\, dj}{\int_{-\infty}^{\infty} e^{-j^2/\sigma^2}\, dj} = e^{-d^2/4\sigma^2}$,

since the integrals in numerator and denominator represent areas under the same curve, shifted $\frac{d}{2}$ units along the horizontal axis. It is of great interest that the form of $\alpha(d)$ does not change rapidly as u_1 and u_2 depart from normality, and the formula $\alpha(d) = e^{-d^2/4\sigma^2}$ or $\alpha(d) = e^{-d^2/2(\sigma_1{}^2+\sigma_2{}^2)}$ holds approximately for quite general u_1 with standard deviation of σ_1 and u_2 with standard deviation σ_2. We prove this now. The area under u_{1j} or u_{2j} need not equal 1, but we multiply u_{1j} by a suitably chosen ε_1 to make the area under $\varepsilon_1 u_{1j}$ add to 1 and the area under $\varepsilon_2 u_{2j}$ add to 1. These curves $\varepsilon_1 u_{1j}$ and $\varepsilon_2 u_{2j}$ are then probability densities and we can use a powerful probability argument to give us the form of α as a function of d. If $\varepsilon_1 = \varepsilon_2 = \varepsilon$, then $\alpha = \frac{\sum_j \varepsilon^2 u_{1j} u_{2j}}{\sum_j \varepsilon^2 u_{1j}^2}$. To make the argument more transparent, we assume the curve u_{2j} is exactly u_{1j} shifted a distance d to the right. Then we can write $u_{2j} = u_1(j-d)$. Let v_j be the mirror image of u_{1j} so that $v_j = u_{1(-j)}$ and $u_{1(j-d)} = v_{(d-j)}$. In these terms, the numerator of α becomes $\sum_j \varepsilon^2 u_{1j} v_{(d-j)}$ or, in integral form if the curves are not discrete, $\int_{-\infty}^{\infty} \varepsilon^2 u_{1j} v_{(d-j)}\, dj$. These are called convolutions of the probability distributions of the random variables u and v; their interpretation is as follows: Each is the probability that the sum of the two random variables is d, for j and $d-j$ always add to d and we are summing over all j values, so we are getting all ways they can add to d.

The "central limit theorem" of probability, one of the most celebrated theorems in mathematics, says that the sum of many such independent random variables will have a very nearly normal distribution. The sum of even two, when they themselves are not too different from normal, will be closely approximated by a normal distribution, and, since the mean of a sum of random variables is the sum of their means (both zero in our case) and since the variance of the sum of two independent random variables is the sum of their variances, we can see that the numerator of α, plotted as a function of d, should be a normal distribution of mean zero and with variance $2\sigma^2$ (where σ as in the figure is the standard deviation of the u curves). If u_1 has σ_1 and u_2 has σ_2, then $\sigma_1 + \sigma_2$ replace $2\sigma^2$. The denominator normalizes α so that it is 1 when d is zero, so

$$\alpha(d) = e^{-d^2/4\sigma^2} = e^{-d^2/2(\sigma_1{}^2+\sigma_2{}^2)} \tag{8}$$

at least to a very good approximation (see Fig. 2-5). This result is of very great importance to us. As d grows from zero, α falls slowly at first and then more rapidly, most rapidly of all at $d = \sqrt{2}\,\sigma$, and then

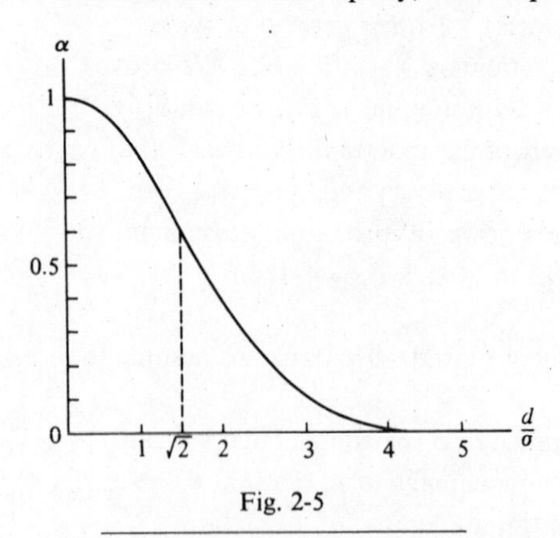

Fig. 2-5

Alpha (α) plotted against $\frac{d}{\sigma}$ according to Eq. (8). Notice that α bows upward for $d<\sqrt{2}\sigma$ and downward for $d>\sqrt{2}\sigma$.

more slowly again. If d is much less than $\sqrt{2}\,\sigma$, α is near 1, while if d is much larger than $\sqrt{2}\,\sigma$, α is near zero. It is around $d = \sqrt{2}\,\sigma$ that α becomes significant. For an empirical case in which d is about equal to σ for a set of birds segregating by vertical foraging height, see Fig. 2-6.

Diffuse competition. There is one final result of great practical significance. We have seen that two nearby competitors are

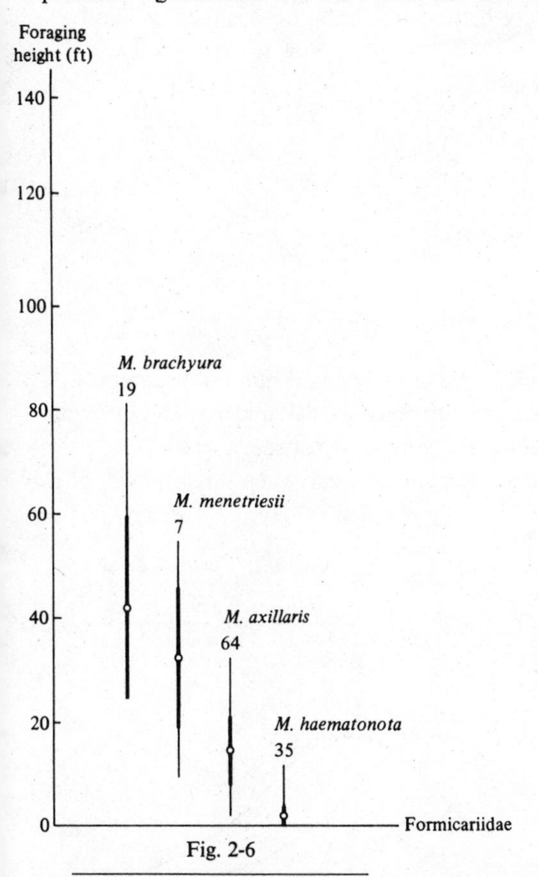

Fig. 2-6

Foraging height relationships within Formicariidae (antbirds). Thickened bars indicate one standard deviation and narrow bars show the entire range of observations. The number of observations of each species of *Myrmotherula* is given over the appropriate bar. Notice that the means are separated by about one standard deviation as the theory indicates. Other species that do not separate by height alone would require more co-ordinates to test the theory. (Provided by J. Terborgh.)

precarious in their coexistence, but coexistence is normally possible no matter how similar the competitors. A species sandwiched between two or even more competitors, however, is in a much worse situation. It has no refuge in either direction, while between two competitors one is always better at something (Fig. 2-7). Here is the theory, in brief, of three species as in the figure, with distance d and competition α between adjacent species. The distance between the outer ones is then $2d$ and by Eq. (8) their competition $\beta = e^{-(2d)^2/4\sigma^2} = \alpha^4$. Now the rest is easy; the species have the equations

$$\frac{1}{X_1}\frac{dX_1}{dt} = \frac{r_1}{K}[K - X_1 - \alpha X_2 - \beta X_3]$$

$$\frac{1}{X_2}\frac{dX_2}{dt} = \frac{r_2}{K_2}[K_2 - \alpha X_1 - X_2 - \alpha X_3] \qquad (9)$$

$$\frac{1}{X_3}\frac{dX_3}{dt} = \frac{r_3}{K}[K - \beta X_1 - \alpha X_2 - X_3]$$

We assume that $K_1 = K_3 = K$ to make the method more transparent; a similar result holds if the K's are different. We ask under what condition species 2 can invade an equilibrium community of species 1 and species 3, for under this condition it can increase when rare and will be a member of the community. Then, by symmetry $X_1 = X_3 = X$, say, and $K = X + \beta X$, whence $X_1 = X_3 = \dfrac{K}{1+\beta}$ at equilibrium. Species 2 can invade this if $K_2 - \alpha X_1 - \alpha X_3 > 0$ or if $K_2 > \dfrac{2\alpha K}{1+\alpha^4}$ or if $\dfrac{K_2}{K} > \dfrac{2\alpha}{1+\alpha^4}$.

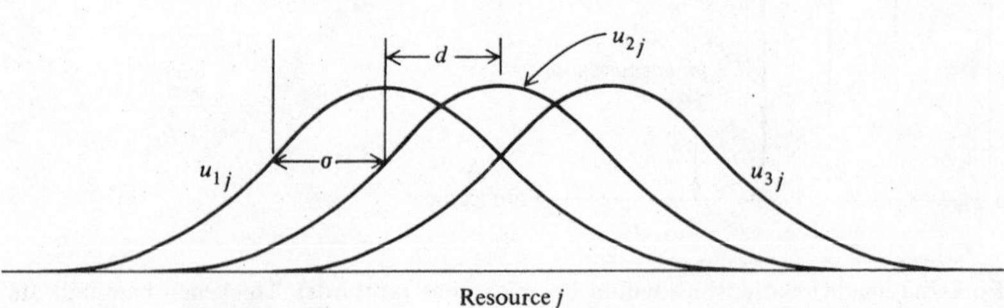

Fig. 2-7

Middle species, with resource utilization u_{2j}, is sandwiched a distance d between two outer ones. See the text for proof that if $d < \sqrt{2}\sigma$ and $K_1 = K_2 = K_3$, the middle species will go extinct. Here $d = \sigma$, so the middle one would go extinct.

Temporarily assume $K_2 = K$ and we have $1 = \frac{2\alpha}{1 + \alpha^4}$, which has the solution $\alpha = 0.544$ corresponding to $d = 1.56\sigma$. Even without climatic variation this middle species is doomed if it is sandwiched in too closely between competitors. Thus a cluster of competitors may much more easily outcompete one species than can a single competitor. It has been suggested that competition by a constellation of species be called "diffuse competition." For example, J. M. Diamond (pers. comm.) has found that in altitudinal sequences of three bird species on New Guinea mountains, the middle species often occupies the narrowest belt and usually is the one that disappears on remote mountains (see Fig. 6-5, p. 139).

There is another way to view the difficulty of the middle of three species. Suppose the distance of the middle from the other species is $D < \sqrt{2}\,\sigma$ (see Fig. 2-8). We consider what would happen if

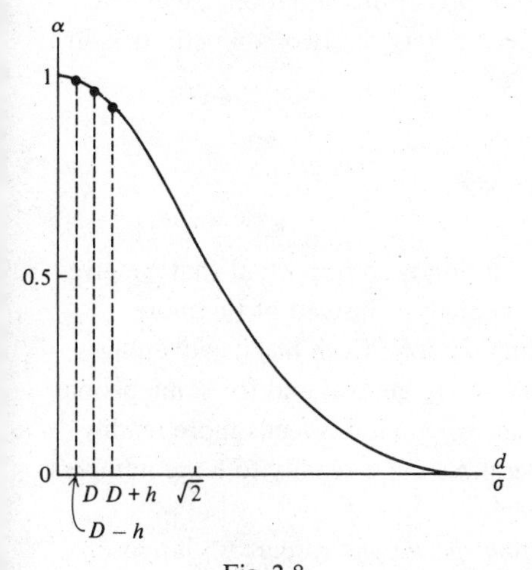

Fig. 2-8

Figure 2-5 with two positions of the middle species of Fig. 2-7 illustrated. If the middle species is distance D from each neighbor, it will have the same competition $\alpha(D)$ with each. If the middle species moves distance h closer to one competitor, it will have competition $\alpha(D - h)$ with the nearer and $\alpha(D + h)$ with the farther. Since the graph shows that $2\alpha(D) > \alpha(D + h) + \alpha(D - h)$, the species will benefit from the convergence. See the text for details.

it moved slightly closer, $(D - h)$, to one competitor and farther, $(D + h)$, from the other. Because the α curve is convex, $\alpha(D - h)$ is closer to $\alpha(D)$ than $\alpha(D + h)$ is to $\alpha(D)$. In other words, the increase in competition from the nearer competitor is less than the reduction of competition with the farther competitor. Thus the middle species benefits by moving closer to one of its competitors; it converges toward whichever is closer. This convergence would occur if $D < \sqrt{2}\,\sigma$, the point of inflection of the α curve. If we repeat the argument with $D > \sqrt{2}\,\sigma$, we see that a shift of the middle species toward one competitor increases competition with the close one by more than it decreases it with the far one. Here the middle species is safest staying in the middle; it would diverge from either competitor to this position. Of course, whether the move is beneficial normally depends upon α, K, and X. These convergence results are imprecise because they fail to incorporate K and X variations, but they give a heuristic view of the reason for convergence. A species so close to a competitor that it must converge and fight it out is clearly in a precarious situation. Again we see that a middle species closely squeezed between two competitors is in deep trouble.

Graphical Treatment of Competition Theory

The theory of competition is so important that we give an alternate treatment using graphical analysis instead of the more mathematical version of the preceding section. Each has its advantages; the graphical analysis is in some ways more general and for some people is easier to understand. The treatment by equations leads more readily to actual numerical results and generalizes more readily to large numbers of competing species.

Numbers of coexisting species and resources. Suppose, to begin with, that we have species 1 and 2 of abundances X_1 and X_2 depending upon resources 1 and 2 of abundances R_1 and R_2. We assume that the consumers are both able to maintain their populations on a mixture of the resources and to increase when the resources are suitably common; conversely, they will decrease if the resources are rare. Thus species 1 will increase outside of some curve such as is drawn in

Fig. 2-9 and will decrease inside it. If the line is straight, we know the resources are interchangeable. A steady population of $R_1 = 6$ and $R_2 = 0$, or of $R_2 = 10$ and $R_1 = 0$, will just support species 1, and, if the resources are interchangeable, we could take half of each—$R_1 = 3$ and $R_2 = 5$—and also just support the population. If the line bows out in the middle, it means it takes more of the mixture than of the pure resources; and if, as seems more probable, the line bows in, it means the consumer can get along on less of the mixture, just as a human being needs less of a balanced diet than an unbalanced one. A line bowing in implies that the resources supply complementary nutrients. A similar line can be drawn for consumer 2, which increases outside of a different line: It can get along on less of R_2 and needs more R_1 if R_1 is eaten alone, and it is hence more of a specialist on the second resource. A glance at the figure shows that the two lines intersect in only one point $R_1 = a$, $R_2 = b$. Biologically this means that X_1 and X_2 are simultaneously able to maintain their populations without increase or decrease only when the resources are at this particular level $R_1 = a$, $R_2 = b$.

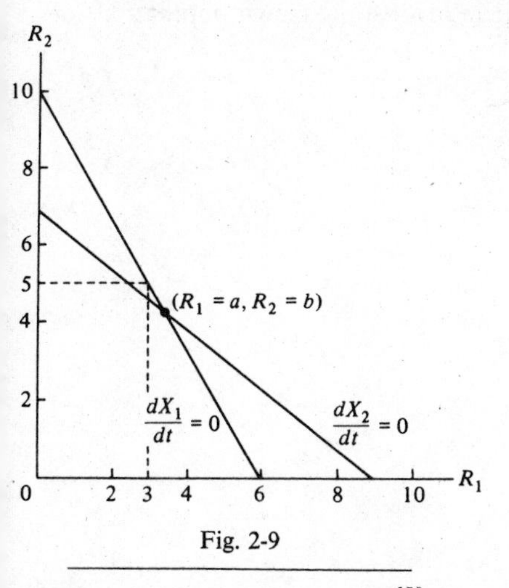

Fig. 2-9

Isoclines for two species. The line $\frac{dX_1}{dt} = 0$ bounds the set of resource values for which species 1 can increase; the line $\frac{dX_2}{dt} = 0$ is the inner boundary of X_2 increase. The lines are drawn straight, reflecting an assumed interchangeability of resources. See the text for a discussion of how the lines would bend if the resources were not interchangeable.

We can immediately draw a few interesting conclusions. First, if we added a third consumer species, its line (or "isocline") separating region of increase from region of decrease would usually not pass precisely through $R_1 = a$, $R_2 = b$ (see Fig. 2-10). This means that each pair of lines has a separate intersection point, each determining unique populations of R_1 and R_2 that will allow those two consumer species to be maintained at equilibrium. Species 1 and 2 remain constant in population if $R_1 = a$ and $R_2 = b$. But at those resource levels species 3 either increases (if its line lies inside a, b) or decreases (if its line falls outside a, b) so this third species cannot be at equilibrium with the first two. Even in the highly improbable case that all three lines actually pass through the same point, it turns out that one is more vulnerable to small population fluctuations and soon loses out. Hence we can have no more than two species coexisting if limited by two resources. Generalizing, no more than n species can coexist if they are limited by only n resources. This interesting statement has two weaknesses. First, if other limiting factors enter—in other words if the equations of growth of the consumers involve their own populations or

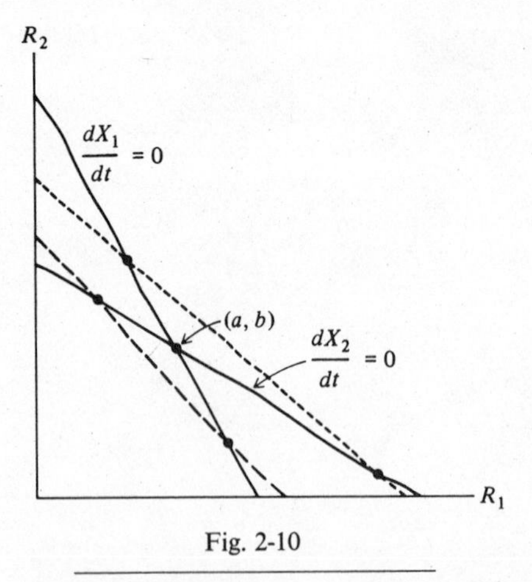

Fig. 2-10

Isoclines as in Fig. 2-9, except that they are not necessarily straight since we are making no assumption about the interchangeability of resources. A third species with the dashed isocline could still increase and hence reduce the resources further when X_1 and X_2 were at equilibrium with $R_1 = a$ and $R_2 = b$. A third species with the dotted isocline that lies outside of (a, b) could not invade the community containing X_1 and X_2 at equilibrium.

predator populations as well as the resource levels—it no longer holds; we could then have more species than resources. Second, no one knows precisely what a "resource" is. It is not a species, because the fruit and bark of a tree can be harvested almost independently by two kinds of consumer insects and hence probably constitute two resources. And the various grass species eaten indiscriminately by some grazers may only constitute one resource. To the extent that we have to tailor our definition of resource to match the species present, we have made no progress.

There is another, less debatable, virtue of Fig. 2-9. We know that the two consumers can coexist at equilibrium only if the resource levels are $R_1 = a$, $R_2 = b$. What if an environment simply would not hold b units of resource 2? Then we are realistically confined to the part of the figure below the line $R_2 = b$, and in that region consumer 1's line is inside consumer 2's line. In other words, consumer 1 can lower the resource to a lower level than can consumer 2 and hence ousts consumer 2 by competition. Species 1 literally lowers the food supply too far for species 2 to tolerate. Conversely, if the environment cannot hold a units of resource 1, then species 2 will outcompete species 1 (see the figure). To give an elementary but hypothetical example, we can picture R_1 as the amount of grass and R_2 as the amount of bushes in a field. Consumer 1 is a good grazer but a poor browser, so is more efficient on grass; consumer 2 is a better browser and poorer grazer, so is better on the bushes. These species can coexist if there is a suitable, precise mixture $R_1 = a$, $R_2 = b$ of the resources, but if the field gets so overgrown with bushes that there is room for less than a units of grass, the browser takes over completely. This shows that, in general, we expect competitors to coexist in some habitats but each to be superior in others. In a very unproductive series of environments, in which maximum quantities of R_1 and R_2 (the upper right corners of the stippled square of Fig. 2-11) formed the dashes along the line of Fig. 2-12, a slightly different result would hold. Instead of a zone of overlap, there would be a zone of species 2 alone, a zone of species 1 alone, and an intermediate zone with neither species. Figure 6-3 (p. 136) shows the altitudinal distributions of two species of warblers (*Myioborus miniatus* and *M. melanocephalus*) and of other species on Peru mountainsides as observed by Terborgh (1971). He observed many congeneric pairs and even triples of species of this sort, and the gaps may possibly be explained in this way.

Evolutionary convergence and divergence. We can also understand evolutionary convergence and divergence in terms of modifications of these graphs (Fig. 2-13). Each phenotype will have its own line $\frac{dX}{dt} = 0$ and R_1 and R_2 intercepts, and the R_1 intercepts of the continuum of phenotypes will form a U-shaped curve as in the figure. The lowest point is the phenotype S_1 (for specialist 1) that can reduce R_1 to the lowest level and still maintain its own population by eating those R_1. As we depart in either direction from this phenotype best able to eat R_1, the R_1 intercepts of these neighboring phenotypes become progressively larger. Similarly, there is an R_2 intercept for each phenotype, but the phenotype S_2 (for specialist 2) best adapted for harvesting R_2 is a different phenotype, so the curve of R_2 intercepts is displaced from the curve of R_1 intercepts. The more different the resources, the more different the phenotypes best equipped to eat them and the farther

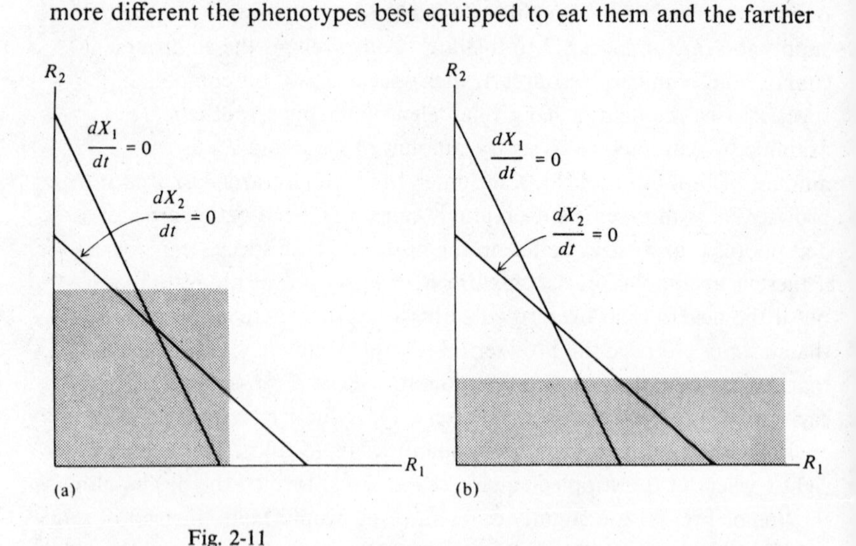

Fig. 2-11

(a) Both resources can reach the same level and the shaded zone includes the intersection of the isoclines for species 1 and 2. Hence both 1 and 2 can persist. (b) Resource 2 reaches only a very low level, perhaps because it is confined to a small part of the environment. Hence, species 1 and 2 cannot both persist; species 1 can always oust species 2 and prevent its reinvasion. (From MacArthur, 1968.)

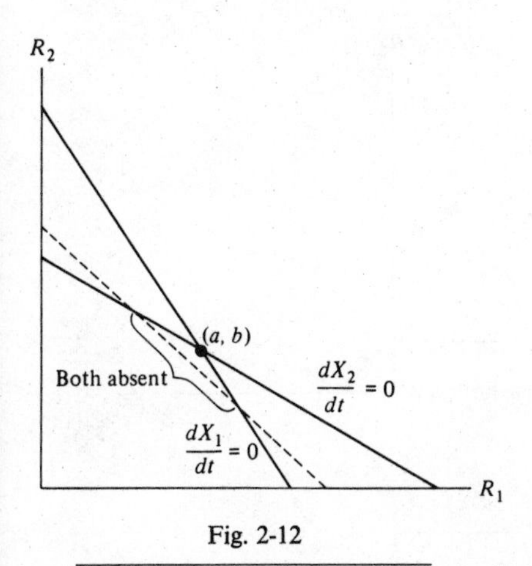

Fig. 2-12

Each dash is the maximum possible R_1 and R_2 in a single habitat. The line of dashes shows a sequence of habitats along a gradient from one that will support no R_1 to one that will support no R_2. The line of dashes lies inside the point (a, b) and intermediate dashes (i.e., intermediate habitats on the gradient) lie inside both isoclines. These habitats will contain neither species. A more productive gradient would form a sequence of dashes lying outside (a, b). Along this gradient we would find first one species alone, then both, then the other species alone.

apart the intercept curves. We can now combine these curves with our earlier graphs by joining the intercepts with straight lines. We inquire about the fate of mutant phenotypes S_1' and S_2', closer together than S_1 and S_2. From Fig. 2-13(a) we prove that when the resources are similar, S_1' and S_2' can invade an equilibrium population of S_1 and S_2 and replace them! In other words, when the resources are similar, evolution favors convergence of specialist phenotypes. In Part (b) of the figure we see that when the resources are very different S_1' and S_2' cannot replace S_1 and S_2 but rather the contrary: S_1 and S_2 could replace S_1' and S_2'. Here divergence to the point of specialization is favored. In the first case, the two resources were too similar to hold two coexisting species such as S_1 and S_2 and have them persist long enough for evolution to take place.

Necessary and sufficient conditions for coexistence. We may proceed to another stage in the theory with a new part for our graph. We have seen that the two consumers can only coexist at equilibrium if the resources are at $R_1 = a$ and $R_2 = b$. But we don't know when the

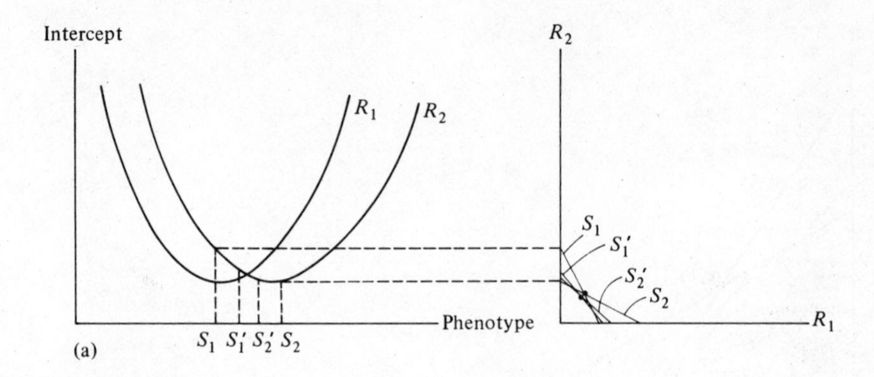

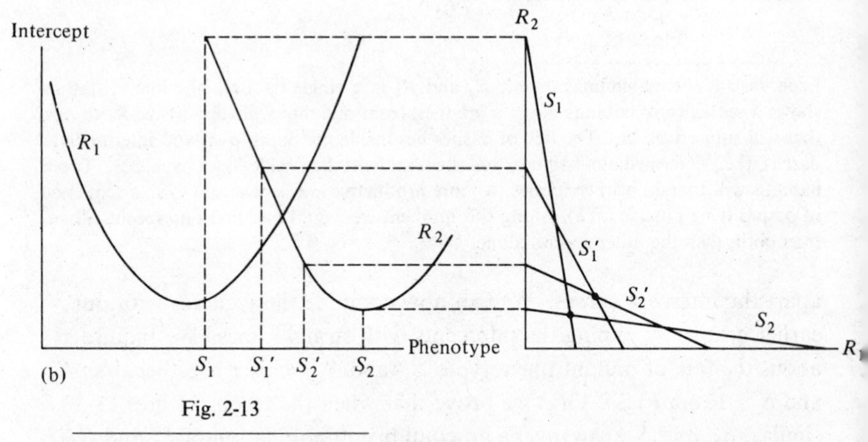

Fig. 2-13

(a) Competition between populations that eat very similar resources (nearby intercept curves). The graph on the right, like Fig. 2-10, proves that convergent phenotypes S_1' and S_2' will reduce the resources more and hence outcompete the specialist phenotypes S_1 and S_2. (b) The resources differ by more (the intercept curves are displaced farther apart), and the graph on the right proves that the convergent phenotypes S_1' and S_2' cannot invade S_1 and S_2. Rather, divergence toward S_1 and S_2 is favored. See the text for details.

resources will reach those levels. We now proceed to give a sufficient condition for equilibrium. The resources will maintain their populations at $R_1 = a$ and $R_2 = b$ only if their increase $\left(\frac{dR_1}{dt}, \frac{dR_2}{dt}\right)$ in the absence of consumers is exactly harvested by the consumers. At this point we must be more precise. Suppose the consumer equations are like those of

Eqs. (3), (p. 37):

$$\frac{1}{X_1}\frac{dX_1}{dt} = C_1[a_{11}w_1R_1 + a_{12}w_2R_2 - T_1]$$

$$\frac{1}{X_2}\frac{dX_2}{dt} = C_2[a_{21}w_1R_1 + a_{22}w_2R_2 - T_2]$$

Then the lines of Fig. 2-9 are the straight lines $a_{11}w_1R_1 + a_{12}w_2R_2 = T_1$ and $a_{21}w_1R_1 + a_{22}w_2R_2 = T_2$. The simultaneous solution of these we called $R_1 = a$; $R_2 = b$. The resources themselves are populations, and we assume that without predation they have a certain rate of increase for each population size. In particular, without predation we assume $\frac{1}{R_1}\frac{dR_1}{dt}$ takes the value $\mathscr{R}_1$, and $\frac{1}{R_2}\frac{dR_2}{dt}$ takes the value $\mathscr{R}_2$, when $R_1 = a$ and $R_2 = b$. These numbers $\mathscr{R}_1$ and $\mathscr{R}_2$ thus measure how, in the absence of predators, the resource populations would move if they started at the intersection point of the lines in the figure. We can represent this with an arrow (Fig. 2-14). Each environment will dictate

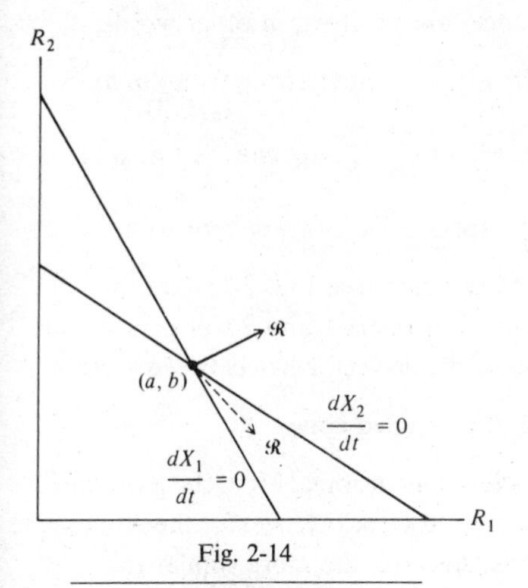

Fig. 2-14

The vector $\mathscr{R}$, with components $(\mathscr{R}_1, \mathscr{R}_2)$, superimposed on Fig. 2-9 with the tail of the vector at the intersection point (a, b). This vector shows how resources at level (a, b) would change in the absence of predation. The solid $\mathscr{R}$ vector shows an environment where both resources would increase when released from predation. The dashed arrow is of the kind that might be observed if the habitat or the season were very poor for R_1. It indicates that when R_1 is at level a with R_2 at b, R_1 must decrease.

the direction of this arrow, and in a seasonably variable environment the arrow may move about. Now the resource growth equations *with predation* are easy to write,

$$\frac{1}{R_1}\frac{dR_1}{dt} = \mathcal{R}_1 - a_{11}X_1 - a_{21}X_2$$

$$\frac{1}{R_2}\frac{dR_2}{dt} = \mathcal{R}_2 - a_{12}X_1 - a_{22}X_2$$

and the two resources are at equilibrium only if both sides are zero,

$$\mathcal{R}_1 = a_{11}X_1 + a_{21}X_2$$
$$\mathcal{R}_2 = a_{12}X_1 + a_{22}X_2$$

or, in vector form,

$$\begin{pmatrix}\mathcal{R}_1\\ \mathcal{R}_2\end{pmatrix} = \begin{pmatrix}a_{11}\\ a_{12}\end{pmatrix} X_1 + \begin{pmatrix}a_{21}\\ a_{22}\end{pmatrix} X_2$$

Now we can say whether both consumer species will in fact be present. They will be present if their abundances are positive; in other words, if the vector $\begin{pmatrix}\mathcal{R}_1\\ \mathcal{R}_2\end{pmatrix}$, which we have already drawn departing from (a, b), is a positive linear combination of the vector $\begin{pmatrix}a_{11}\\ a_{12}\end{pmatrix}$—species 1's foraging probabilities—and the vector $\begin{pmatrix}a_{21}\\ a_{22}\end{pmatrix}$—species 2's foraging probabilities. So now we add these two vectors to the figure (see Fig. 2-15), putting their tails on (a, b), and we know that consumers 1 and 2 will persist with positive abundances if and only if the $\mathcal{R}$ vector lies between these foraging vectors $\begin{pmatrix}a_{11}\\ a_{12}\end{pmatrix}$ and $\begin{pmatrix}a_{21}\\ a_{22}\end{pmatrix}$ in the stippled zone.

We can now draw a few conclusions. First, the more variable the environment, the more precarious the coexistence, for the $\mathcal{R}$ vector may wander outside the stippled zone. Second, the more similar the species, the more precarious the coexistence. This is seen most directly in the case where both resources have the same weight so that $w_1 = w_2$. Then the vector $\begin{pmatrix}a_{11}\\ a_{12}\end{pmatrix}$ is perpendicular to the line of species 1, which now

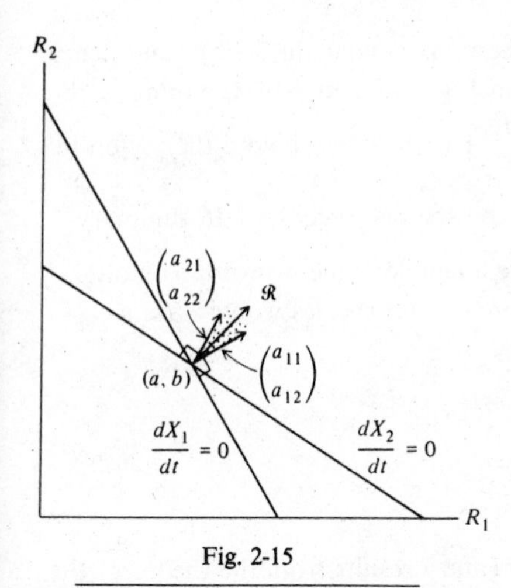

Fig. 2-15

Figure 2-14 with the utilization vectors $\begin{pmatrix} a_{11} \\ a_{12} \end{pmatrix}$ of species 1 and $\begin{pmatrix} a_{21} \\ a_{22} \end{pmatrix}$ of species 2 added. When $\mathscr{R}$ lies in the stippled region between utilization vectors, X_1 and X_2 are both positive at equilibrium. If $\mathscr{R}$ lies outside the stippled zone, either permanently or seasonally, one species vanishes. See the text for details.

takes the form $w(a_{11}R_1 + a_{12}R_2) = T_1$ or $a_{11}R_1 + a_{12}R_2 = \frac{T_1}{w}$. To see this, let (R_1', R_2') be some arbitrary second point on this line, with (R_1, R_2) being an arbitrary first point. Thus both $a_{11}R_1 + a_{12}R_2 = \frac{T_1}{w}$ and $a_{11}R_1' + a_{12}R_2' = \frac{T_1}{w}$. Subtracting, $a_{11}(R_1 - R_1') + a_{12}(R_2 - R_2') = 0$, which can be written in vector form $\begin{pmatrix} a_{11} \\ a_{12} \end{pmatrix} \cdot \begin{pmatrix} R_1 - R_1' \\ R_2 - R_2' \end{pmatrix} = 0$, which shows that the inner product of these vectors is zero, meaning $\begin{pmatrix} a_{11} \\ a_{12} \end{pmatrix}$ is perpendicular to the vector $\begin{pmatrix} R_1 - R_1' \\ R_2 - R_2' \end{pmatrix}$, which lies along the line bounding consumer 1's zone of increase. $\begin{pmatrix} a_{11} \\ a_{12} \end{pmatrix}$ is thus drawn perpendicular to the first line in the figure and $\begin{pmatrix} a_{21} \\ a_{22} \end{pmatrix}$ perpendicular to the second line. Now we see that the more similar the species and hence the more nearly parallel their lines in Figs. 2-9 and 2-14, the more nearly parallel are the $\begin{pmatrix} a_{11} \\ a_{12} \end{pmatrix}$ and $\begin{pmatrix} a_{21} \\ a_{22} \end{pmatrix}$ vectors that bound the region in which the $\begin{pmatrix} \mathcal{R}_1 \\ \mathcal{R}_2 \end{pmatrix}$ vector must lie for the species to coexist. In summary, more similar competing species have a smaller zone of overlap. Note that if they were not competing, more similar species would have a larger zone of overlap.

The Equilibria of Predator-Prey Interactions

There are two elementary results from the theory of the statics of predation that are relevant to biogeography and will be described here. We consider a consumer whose population obeys the familiar equation $\frac{1}{X}\frac{dX}{dt} = C(aR - T)$ and, to begin with, a resource growing according to $\frac{1}{R}\frac{dR}{dt} = \frac{r}{K}(K - R) - aX$. Here as before (p. 37), R is the amount of resource present, a is a probability of an X individual catching and eating an R in a unit of time, and T is the number needed to support the X population. Thus $aR - T$ is the food present in excess of that needed for maintenance. This excess can be converted into the per capita rate of increase, $\frac{1}{X}\frac{dX}{dt}$. The second equation shows that, in the absence of predators, X, the per capita rate of increase of R, $\frac{1}{R}\frac{dR}{dt}$, is logistic; i.e., it takes the value r when R is small and falls to zero as R approaches K. In the presence of predation, each of the X predators removes fraction a of the resource individuals per unit of time, where a is the same as in the first equation. At equilibrium the right sides of these equations are zero, whence the equilibrium values of X and R are given by $X_e = \frac{r}{a}\left[1 - \frac{T}{aK}\right]$; $R_e = \frac{T}{a}$. First, we notice that X can over-exploit its resource by increasing its a too much. In fact, setting $0 = \frac{\partial X_e}{\partial a}$

we find X_e is a maximum when $a = \frac{2T}{K}$, and X_e declines for larger a. Second, competition favors large a. In fact, a second species with larger a will cause smaller $\frac{T}{a}$ and will thus reduce the resource equilibrium to a lower level and hence outcompete the first species. We conclude that competition can lead to overexploitation of this kind of resource. A resource like falling fruit or emerging insects, whose rate of appearance is independent of the rate at which it is harvested, will obey a different equation, such as $\frac{dR}{dt} = F - aXR$. Now, in the absence of predation the resource accumulates at the constant rate, F, of fall of fruit. Each of the R fruits has probability a of being eaten by any given predator, so each predator eats aR fruits per unit time and the X predators in combination eat aRX. Hence $X_e = \frac{F}{aR}$ at equilibrium when $\frac{dR}{dt} = 0$. As before $R_e = \frac{T}{a}$ so $X_e = \frac{F}{T}$. Such a resource cannot be overexploited because X_e is independent of a. In conclusion, some, but not all, kinds of resources can be overexploited and competition may lead to such overexploitation. The testable aspect of this result is that an island with inappropriate species (i.e., smaller a) could maintain a larger population of these species than the mainland could of its better competitors; that is, total population densities can be greater on islands. This phenomenon certainly exists (see p. 112), but there are other possible explanations.

The second result of biogeographic interest is a consequence of the same equations. In particular we saw $R_e = \frac{T}{a}$: an efficient predator with large a keeps the resource level low. If there are many resources, R_i, the population growth of a predator who switches to the commonest resource could be described by $\frac{1}{X}\frac{dX}{dt} =$ $C\left(a \max_i R_i - T\right)$ where $\max_i R_i$ stands for the largest of the R_i for all i, whence the largest R_i and therefore all R_i, would be kept lower than $\frac{T}{a}$. If all resources are kept this scarce, and if $\frac{T}{a}$ times the number of resource species is not enough to fill the environment, there is room for more resource species. This is a form of Janzen's explanation of why there are so many tree species in the tropics (see p. 191). A single predator who does not switch or have a search image would not have this effect, although separate predators, one for each resource, would also set ceilings on the abundance of each resource.

[Ed. note: There is some ambiguity of notation in the equations on pages 38 and 39. The K_1 and K_2 on the left-hand sides of Eq. (7) refer to the "carrying capacities" for the competitors in Eq. (1); the K_j on the right-hand sides of Eq. (7) refer to the resource renewal terms in Eq. (4).]

The Economics of Consumer Choice

3

Where should an animal feed to get the most food, and what items of food should it pursue? To answer these questions some knowledge of the structure of the environment is essential since what is optimal foraging in one environment need not be in another. Some knowledge also about the animal's own foraging apparatus is needed since what is optimal for a woodpecker is not optimal for a heron. Thus from three ingredients, structure of the environment, functional morphology of the foraging species, and knowledge of its economic goals, we can draw conclusions of interest. The treatment here is an improved version of that given by MacArthur and Pianka (1966). Schoener (1969) has given an alternate form to what is essentially the same theory, but his paper contains other applications of the theory.

Preliminary Assumptions

First, we consider the structure of the environment: The minimum assumption seems to be that of repeatability. We assume the diversity of environments is repeated so that a consumer searching in a second patch can encounter the same spectrum of resources that it did in the first. Thus, from a foraging bird's point of view, there is a great diversity of foliage types, from tree trunks to loose canopy to bushes to herbaceous ground cover, but each is repeated, so that a tree trunk feeder, for example, can find many trunks to search. This postulate at least is required because it means that a searching bird will have a fairly clear statistical expectation of the resources it will come upon. Without this—and it appears reasonable—we can make little progress in understanding the economics of the choice of resources worth pursuing.

It will sometimes also be convenient to make a more stringent assumption, which holds only for certain foragers. We assume the spectrum of resources is continuous and unimodal. Thus the insects eaten by birds in the summer come in a continuum of sizes and most are of intermediate size. Hespenheide (1971) and Schoener and Janzen (1968) have shown that bird food comes in a log-normal distribution of sizes (Fig. 3-1); the exact shape is of less consequence to us. (This assumption clearly would not hold for the food of monophagous insects or of herbivores in general, since the chemistry of plants seems often to come in discrete categories.) Both of our assumptions are for the purpose of reducing chaos.

We turn next to the assumptions about functional morphology, and we use only the most rudimentary ones because knowledge is so meager. Some sort of continuity principle seems to be minimal: that nearby morphologies are adapted for nearby methods of harvesting food. This also implies that if one species is intermediate in morphology between two others, it will, as a rule, be adapted for an intermediate resource supply. This applies best to species that are similar to begin with. Thus three woodpeckers differing virtually only in size—one large, one medium, and one small—would be expected to have correspondingly large, medium, and small food sizes and perch diameters. However, if we could imagine a bird precisely intermediate between a woodpecker and a swan—birds already very different—it would be dangerous to infer its food supply.

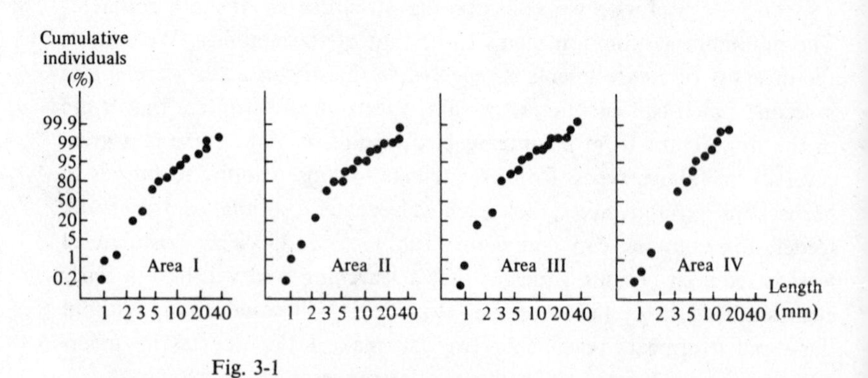

Fig. 3-1

Distributions of insect length plotted on logarithmic probability paper. From four areas in Costa Rica. Area I is lowland tropical dry forest; III is river-bottom forest; II is the edge between I and III; IV is lowland tropical rain forest. (From Schoener and Janzen, 1968.)

This continuity has an inverse aspect: that nearby foods are gathered with nearby efficiencies. Thus if a flycatcher is best adapted for catching beetles 10 mm long, its efficiency at 9 mm and 11 mm should be just slightly less, and so on.

There is a yet more profound assumption having to do with the perfectability of tools. In human affairs we express it by saying "a jack of all trades is a master of none." It tells us that a harvester cannot be simultaneously perfect at several jobs; perfection in one involves reduced efficiency in another, and if an organism must try to harvest in various ways, it must compromise its efficiency in each. But since competition often puts a premium on efficiency, this assumption implies a division of labor among specialists. It is the ultimate reason we have so many species.

We are able now to draw some conclusions about consumer choice. Although the assumptions we have made so far seem quite innocuous, some nontrivial results can already be derived. We must assume some goal of the species' behavior, and since we have been talking about resources, we shall assume that the goal of our species is to maximize their intake of these resources.

Optimal Feeding Place and Diet

The gathering of food is divided into four phases: First, deciding where to search; second, the search, keeping a lookout for various kinds of palatable items; third, on locating a potential food item, the decision whether to pursue it; fourth, the pursuit, with possible capture and eating. By our assumptions about morphology and the repeatability of the environment, the efficiencies of search and of pursuit are determined for each kind of resource and only the correct decisions for phases 1 and 3 need to be determined. The first decision—where to feed—is easy: the species should forage where the expectation of yield is greatest. This foraging place need not remain constant. Yields may vary from season to season, with the species adjusting its foraging place accordingly. More drastically, food densities may be greatly altered by competitors. Hence on islands, with reduced competition, species usually forage in different locations and over a wider range than on the mainland.

There is a simple criterion for the phase 3 decision as to which located items to pursue. On locating an item, the consumer has two choices: to pursue or to search again for a better item and pursue it instead. Since either choice ends with the animal ready to start a new search, the better decision is the one that promises to yield more per unit of time. In other words, an animal should elect to pursue an item if and only if, during the time the pursuit would take, it could not expect both to locate and to catch a better item. Stated more formally, if P_j is the pursuit time per gram successfully captured for the just located jth kind of resource and $\bar{P}$ and $\bar{S}$ are the average pursuit and search times (per gram successfully captured) for the items in the previous diet, item j should be added to the diet if and only if $P_j \leqslant \bar{P} + \bar{S}$. This can either be viewed as intuitively true or can be derived from the requirement that the diet should be so chosen that $\bar{P} + \bar{S}$ is minimum. In either case we require a repeatable environment so that the likelihood of a better item can be assessed rationally.

Applications

The criterion for the phase 3 decision is easy to apply to some situations. First let us suppose that, as with foliage gleaning birds, the search is always moderately time consuming, but the pursuit, capture, and ingestion of the minute, stationary insects take negligible time and are almost always successful. Then $\bar{S}$ is large compared to $\bar{P}$ and all such stationary insects that are palatable should be part of the diet—for always in such cases $P_j < \bar{S} + \bar{P}$. Thus "searchers" should be generalists. Conversely, a "pursuer"—a species with prey always in sight, so that $\bar{S}$ is negligible, but for whom pursuing is a time-consuming business (perhaps a lion)—should always select the prey with the smallest P_j. Similarly, if we contrast an unproductive environment (large $\bar{S}$) with a productive one (small $\bar{S}$), we can infer that the same species should be more specialized in the productive than the unproductive environment. Thus H. F. Recher (pers. comm.) found that great blue herons (*Ardea cinerea*) in productive Florida waters selected a much narrower range of food size than they ate in the unproductive lakes of the Adirondacks.

From the present viewpoint, the most important thing to note is that the rarity of a food has very little to do with its being included in, or excluded from, the diet; for P_j, the pursuit time, is essentially independent of abundance (although the irrelevant S_j depends strongly upon it). The general level of food abundance, which controls $\bar{S}$, is important, but the rarity of a particular item has no effect upon its inclusion. Thus if the environment contains a continuum of food types, and if the species' morphology is efficient at catching a certain segment of the continuum, we expect the species to accept a segment as its diet. Of course, unpalatable foods are excluded.

Mental limitations of the forager may force a slightly different pattern of foraging. A person may be able to search a random list of surnames for Smith and Jones simultaneously, but if he is searching for 25 surnames, he may fall into a pattern of searching the list first for one group and then the same list for a second, and so on. When a bird does this, we say he has formed a "search image," whereby he may at first overlook palatable beetles in his search for green caterpillars. Such a search image will still be proper economic behavior either if the green caterpillars are in a slightly different location, so that it is possible not to encounter the beetles (not really a search image in the true sense), or the bird is incapable of simultaneously remembering that both beetles and green caterpillars are good to eat, or if the struggle to remember hampers his search.

Cody (1971) has studied the economics of the flocking behavior of fringillid species in the Mohave Desert. He argues that foraging flocks accomplish two things that are not readily accomplished by an individual acting independently. First, exploited vs. unexploited areas are more easily recognized, with the result that $\bar{S}$ is reduced. Second, in the case of renewable resources, flock behavior can evolve in such a way as to optimize the return time to previously exploited areas.

There are many more conclusions of importance to be drawn from our discussion of consumer choice. First we infer what has been called the "compression hypothesis" (MacArthur and Wilson, 1967). This hypothesis states that as the number of competing species increases, feeding habitats contract, and the actual range of food items taken either remains constant or actually increases slightly (Fig. 3-2). To support the first part of the hypothesis we recall that the presence of competitors will

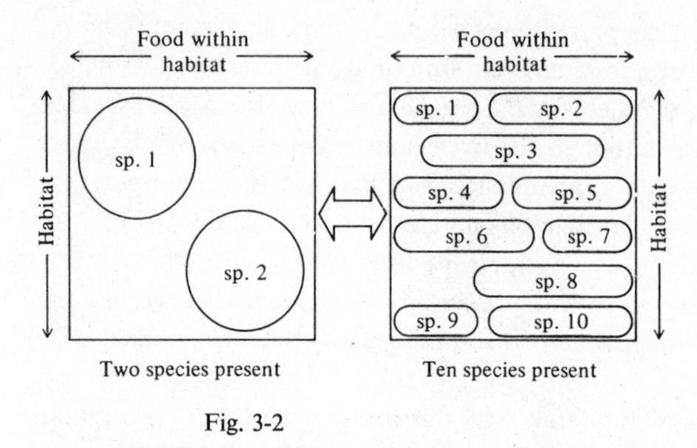

Fig. 3-2

The compression hypothesis. As more species invade and are packed in, the occupied habitat shrinks (although some marginal patches may be added) but not the range of acceptable food items within the occupied habitat. The actual diet, reflecting both acceptability and abundance of items, may become more concentrated, but the range of items should not greatly change. Conversely, as species invade a species-poor island from a species-packed source, only the occupied habitat expands. The hypothesis applies only to short-term, nonevolutionary changes. (From MacArthur and Wilson, 1967.)

reduce the food supply, and hence the expected harvest, in zones where the competitors feed. Therefore, the selection of feeding places should normally be restricted by competitors in accord with the compression hypothesis. Occasionally, however, the competitors may reduce the food in a species' own favored feeding location. In such situations the effect would be to increase the species' range of foraging places.

To support the second part of the hypothesis we remember that, within the zone selected in which to feed, rarity plays little part in the inclusion of a food item in the diet, so that if a competitor reduces a potential food item, it will not thereby affect the inclusion of the item in the diet of another species. However, if the competitor reduces the density of a whole segment of the diet, this materially increases mean search times and may change pursuit times also, which may affect—normally increasing—the range of food items taken. Accordingly, neither aspect of the compression hypothesis is airtight, but both should apply commonly. The habitat compression has been documented often (pp. 85-89; 135-140) but range of foods taken is much more difficult to measure. Diamond (1970b) recorded that of 52 New Guinea birds on the island of Karkar, 29 underwent habitat expansion, yet only one expanded its diet.

For most species small foods are common while large and very large foods are quite rare. This means that a small forager, of a size appropriate for eating the small foods, will have a small search time, S, and hence a fairly restricted diet. A much larger forager of the same type will feed on the rarer end of the food spectrum, where S is large, and hence must eat a wider range of foods. This is easy to document (Fig. 3-3). A comparison of Fig. 3-3 with Figs. 3-4 and 3-5 shows that in plotting food sizes it is better to plot logarithms of size, which tends to make the curves symmetrical and of equal width. This in turn explains why spacings between consumer species' sizes are uniform on a logarithmic scale; or in other words, why there tends to be a constant difference between logarithms of size. Finally we note that constant difference between logarithms implies a constant ratio of sizes: $C = \log A - \log B = \log \frac{A}{B}$, whence $\frac{A}{B}$ is constant. In other words, we get a glimpse of why large species are a constant multiple of the sizes of smaller species rather than showing a constant difference in weight. For example, Storer's study (1966) of the three North American accipiters, in which he compiled the mean weights of the male and female of each species, shows that the larger species is roughly 2.5

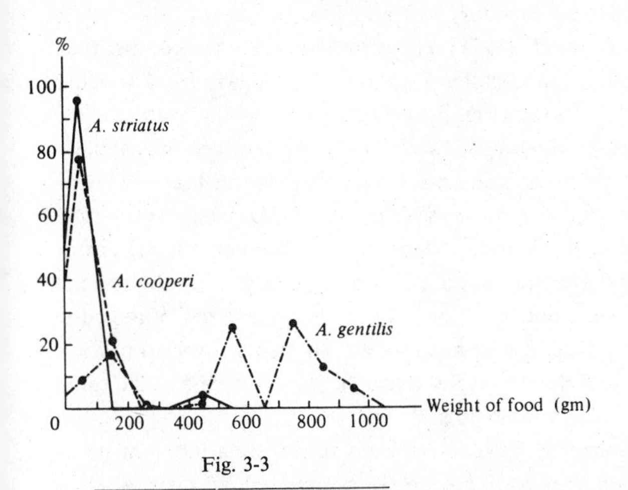

Fig. 3-3

Frequencies of different food sizes taken by the small (♂ = 99 gm, ♀ = 171 gm) sharp-shinned hawk (*Accipter striatus*), the medium-sized (♂ = 295 gm, ♀ = 441 gm) Cooper's hawk (*A. cooperi*), and the large (♂ = 818 gm, ♀ = 1137 gm) goshawk (*A. gentilis*) showing the larger food range of larger species. (Data from Storer, 1966.)

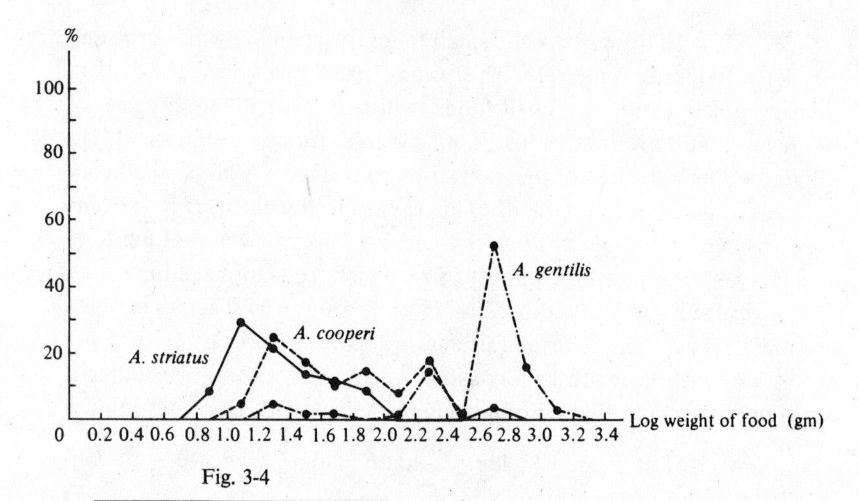

Fig. 3-4

The data shown in Fig. 3-3, but plotted with the logarithm of the food size as the coordinate. Now curves are more symmetrical and of more equal width.

times the weight of the smaller one. Storer listed the mean weights in grams, male and female, respectively, as follows: *Accipiter gentilis*, 818 and 1137; *A. cooperi*, 295 and 441; *A. striatus*, 98.8 and 171.

Hespenheide (1971) relates overlap, or α, the competition coefficient (as defined in the Chapter 2 appendix), to the ratio of weights of larger consumer to smaller (Fig. 3-6). Each point on the graph corresponds to a species pair from one of two insect-eating families—flycatchers and swallows. The vertical axis indicates size similarity in beetles in stomach contents expressed as α. Generally, the larger the weight ratio, the less overlap there is in the length of food items, although a weight ratio of 2 in the flycatchers corresponds to a surprisingly large α of about 0.85. However, weight alone is not the only factor to be considered. Vireos, for example, segregate by habitat and not by size. Thus their overlap in food size is high. On the other hand, several flycatchers may coexist in the same habitat, where usually they differ in size.

Schoener (1969), deriving his results by a different pathway, relates convergence of body size to the economics. We have already remarked that a competitor that reduces the abundance of all food items increases mean search time, $\bar{S}$, and hence causes an enlargement of the optimum diet in another species. When faced with such a competitor, a large

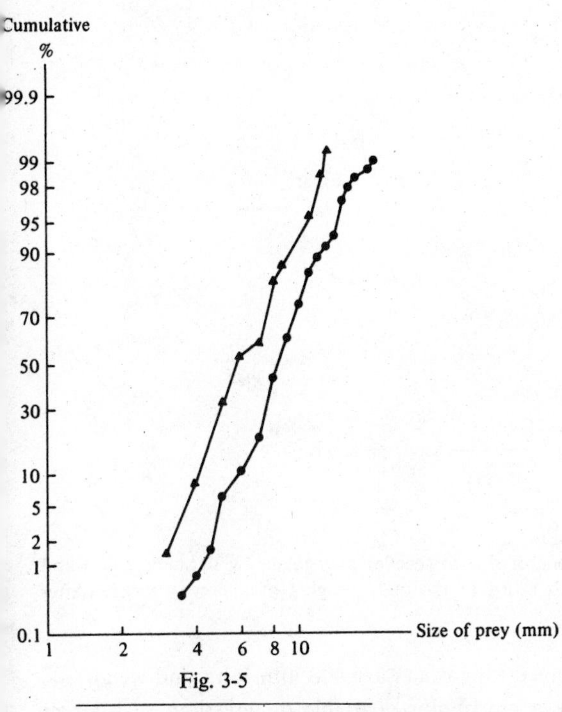

Fig. 3-5

Plot of food sizes of the crested flycatcher (*Myiarchus crinitus*) (●) and the wood pewee (*Contopus virens*) (▲). These are plotted on logarithmic probability paper, so that a straight line corresponds to a log-normal distribution. The mean logarithm of food size of the crested flycatcher is 0.911 and that of the wood pewee is 0.731. Their standard deviations are 0.215 and 0.209, respectively, so they differ by about one standard deviation. (From Hespenheide, 1971.)

species, which is already eating the rare end of the food spectrum, will have to expand its diet, which means taking more small items. Hence, when faced with a generalist competitor or the equivalent in a combination of competitors, a large species may become reduced in size. This is Schoener's conclusion, which he supports by showing how the mean size of lizards of the genus *Anolis* is reduced on islands with many species of competitors (Fig. 3-7).

To show the interaction between feeding position and diet, we now examine the feeding positions of two kingfishers, green (*Chloroceryle americana*) and ringed (*Ceryle torquata*) in Panama. Green kingfishers are small birds with an average length of 182 mm and a weight of

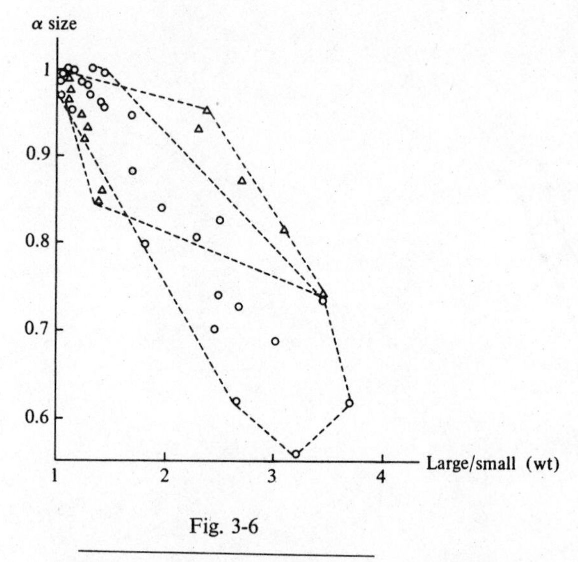

Fig. 3-6

Overlap, or α, in distributions of size of (beetle) prey taken by flycatchers (○) and swallows (△) compared with ratios of the body weights of the two species. (After Hespenheide, 1971.)

38 gm. The very large ringed kingfishers are 400 mm long and weigh 300 gm; they obviously have a morphology capable of wrestling with larger fish than the green can manage, and by virtue of their large weight they must need more grams of fish per day to meet their metabolic needs. The green kingfisher must eat small fish and hence must perch near the water, where the small fish are close enough to be visible. The ringed should perch where the greatest number of grams of fish per day can be captured, so it perches high enough to search a wide area for big fish. But notice how this restricts its diet: by perching so high that it can survey a large area, it can no longer see the very small fish, or if they are visible, the energy it would get by eating one would not compensate for the energy expended in the long dive. Hence the ringed kingfisher is largely confined to eating big fish, and its feeding position has affected its diet.

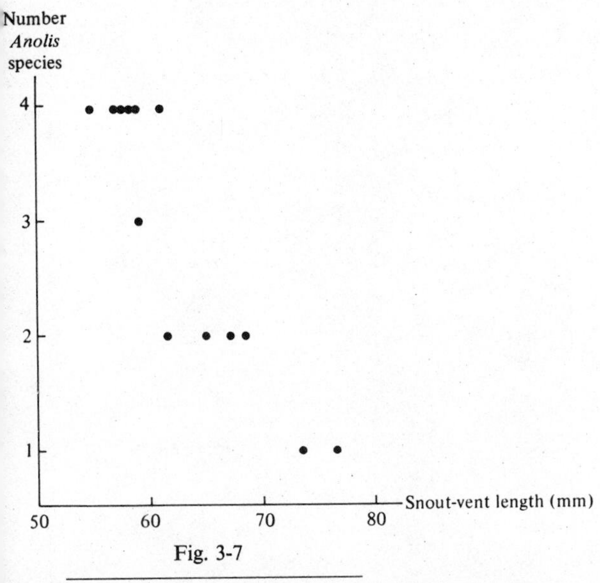

Fig. 3-7

Mean size of lizards plotted against number of species. Where there are more lizard species, their sizes tend to be reduced, as predicted. Each dot represents the mean of the largest third of the sample from a single island. This is chosen to reduce the effect of growth in the lizards: choice of the largest third represents comparison of mature lizards. (After Schoener, 1969.)

The Geography of Species Classification

4

Species are the units, in fact the only units other than the individual organisms themselves, in the organization of a community. A converse statement is also true: Except in a community of coexisting individuals, species may lose their objectivity. At this point, we content ourselves with considering only the first of these. It is not quite true that all organisms of a community are organized into clear-cut objective species, for it takes the cohesion due to sexual reproduction to hold species together. Consequently, among plants and more primitive animals that frequently use asexual reproduction all sorts of intermediates between forms the taxonomist names as species may be found. Poplars and willows are famous for this. But among organisms that always reproduce sexually, species are very well defined (assuming we restrict ourselves to the coexisting individuals we might find at one place and time). The bird censuser who counts and identifies all the species present in 5 or 50 hectares rarely has trouble attaching a species name to any individual.

The picture changes completely when we consider species over a wide geography. No longer is there any cohesion even in sexual forms, for a California robin is not benefitted by resembling a Newfoundland one. To understand the geography of species, we begin by recalling how species form. Except for polyploidy, which is quick but not particularly clever since it involves doubling all genes rather than selecting a useful combination, species form by divergence in geographic isolation followed by a new geographic overlap and coexistence. Once two species so formed are coexisting, the process has gone to its completion. But this process takes thousands of years, and there is every reason that we should expect to see all intermediate stages if we look around us. We should find some populations diverging in isolation but still clearly assignable to the same name on all morphological grounds and hybridizing freely where they may meet if the isolation is not quite perfect. This entire population, isolate plus parent population plus hybrids, is a *polytypic* species, and most species belong in this category. We should also expect to see the next stage, in

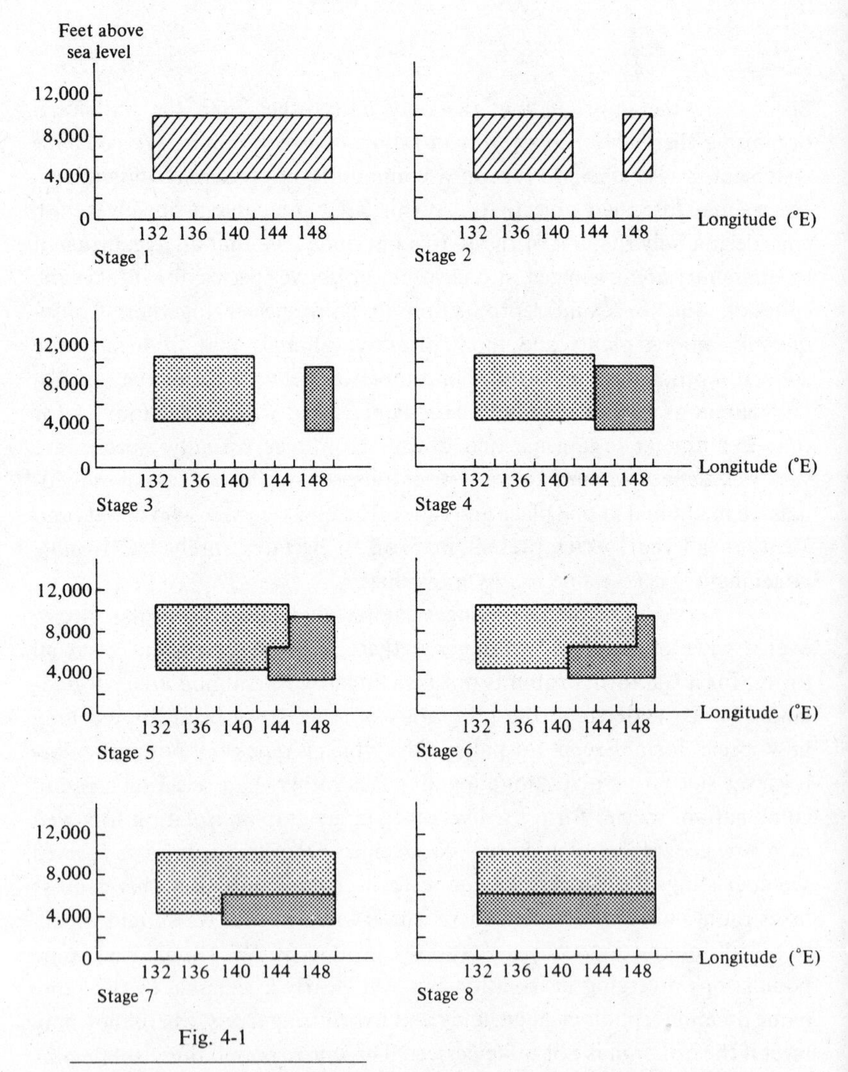

Fig. 4-1

The various stages of species formation in New Guinea. For a description of the stages and a discussion, see the text. (Provided by J. M. Diamond.)

which the isolated population has become sufficiently different to be reproductively isolated from the parent population (we imagine a small zone of contact to test this possibility) but not yet sufficiently different so that they can coexist under competition. Hence there is no overlapping. This whole population is called a *superspecies* and the apparently reproductively isolated components are *semispecies.* Of course, a polytypic species may in time become a superspecies, and there is no sharp line dividing one from the other. For practical purposes, the populations are said to have reached superspecies level if, in the judgment of the systematist, the degree of hybridization is slight enough so that the population's further divergence is not prevented. If the semispecies or superspecies diverge further and are able to coexist in some environments, they clearly become full species, but it is convenient to admit their very close relationship by calling them a *species group.*

Speciation in New Guinea montane birds has been studied by Diamond. The eight presumed stages in the transformation of one species into two species with mutually exclusive altitudinal ranges are illustrated in Fig. 4-1. The evidence for this process cannot be based on direct observations of a single species successively going through these stages, since ornithologists have been studying New Guinea birds for only a century. It is based instead on reconstruction: The distribution patterns of different species or pairs of species can be thought to represent "snapshots" of different stages in a continuous speciation process.

The mountains of New Guinea extend nearly uninterruptedly in an east → west direction from longitude 132°E to 150°E. Forest extends uninterruptedly from sea level to a timberline at about 12,000 feet. Each stage in the figure depicts bird altitudinal distribution as a function of longitude. The stages are as follows:

Stage 1. A single montane species without close relatives extends from the western to the eastern end of New Guinea and occupies approximately the same altitudinal range at all longitudes. Diamond lists 11 examples.

Stage 2. The local population in one area of New Guinea dies out, so that the east-west distribution becomes discontinuous. However, the western and eastern populations are still sufficiently similar to surely interbreed if contact were reestablished; hence they are still considered members of the same species (although possibly different subspecies). Diamond lists 10 examples of this distribution pattern.

Stage 3. In isolation, the eastern and western populations diverge sufficiently so that they would probably not interbreed if contact were established; i.e., they are assumed, without conclusive proof in the form of actual contact and noninterbreeding, to be distinct species (actually, distinct semispecies of the same superspecies). They may also develop slightly different altitudinal ranges. Three pairs are cited as examples.

Stage 4. Both populations reexpand geographically until their geographical ranges abut but do not overlap ("parapatric" distribution, in contrast to stage 3, "allopatric" distribution). There is no or little interbreeding, proving that the populations are in fact distinct species. (In other cases, where divergence in isolation did not proceed far enough to prevent interbreeding on reestablishment of contact, the result is merely intergrading western and eastern subspecies, and there is still only one species.) Diamond notes two pairs as examples.

Stage 5. Each species begins to expand geographically into the range of the other, so that there is overlap ("sympatric" distribution) for a short distance. Within the zone of overlap the two species segregate altitudinally, each being confined to the altitudinal range in which it is competitively superior to the other. This compression of altitudinal range in the zone of sympatry compared to altitudinal range outside the zone of sympatry is a clear demonstration of niche compression due to competition. Two pairs illustrate this.

Stage 6. Both species continue to expand, and the zone of overlap now covers much of central New Guinea. Strict altitudinal segregation in the zone of overlap is maintained. As examples there are five pairs.

Stage 7. As expansion continues, the western species reaches the eastern end of New Guinea and overruns the entire geographical range of its eastern sibling. The eastern species has given up the upper (or lower) part of its altitudinal range throughout its entire geographical range and has not yet reached the western end of New Guinea. Four pairs are cited.

Stage 8. The eastern species continues to expand until it has reached the western end of New Guinea. The two species are now sympatric over the entire length of New Guinea, with mutually exclusive altitudinal ranges. Nine pairs are listed. Diamond notes that in these cases of completed speciation one cannot see which species originated in the west and which in the east.

After stage 8, evolution can proceed in either of two further directions: (a) The process may repeat itself one or two times to yield series of three or four closely related species sympatric over the whole of New Guinea but occupying mutually exclusive altitudinal ranges. Four genera illustrate this. (b) Two species that have become sympatric with mutually exclusive altitudinal ranges may diverge in other niche parameters besides altitudinal preference, so that partial altitudinal overlap becomes possible; for example, as a result of difference in body weight (three species pairs), adaptation to foraging at different vertical heights within the forest (two pairs), or different foraging techniques (two pairs).

The relevance to the biogeographer of the classification we have described in this chapter is quite clear. If he wants to study cases of competitive exclusion, he should look for a zone of contact between semispecies of a superspecies. If he wants to study how similar coexisting species can be, he should study slightly overlapping species of a species group. Mayr and Short (1970) have provided a very convenient book listing the North American bird species and judging to which category they belong. We follow their bird species names in this book.

Part II
The Patterns

The concept of pattern or regularity is central to science. Pattern implies some sort of repetition, and in nature it is usually an imperfect repetition. The existence of the repetition means some prediction is possible—having witnessed an event once, we can partially predict its future course when it repeats itself. The imperfection of the repetition gives us the means of making comparisons. We witness an event A, occurring under conditions C, then, under slightly altered conditions, C′, we witness a slightly altered event, A′. Now we have the seed of a scientific hypothesis: "the difference between C and C′ causes (i.e., is always associated with) the difference between A and A′," which we test by further observations. In geographic ecology, we study patterns repeated in space, not time, and the natural comparisons are those of events occurring in different places. Over and over again in what follows we compare the species on the mainland to those on an island, the species on one mountain to those on another, the species high on a mountain to those lower on the mountain, the communities of the tropics to those of the temperate, and so on.

We begin with an account of islands, their colonization, their loss of species by extinction, and the lessons to be learned by comparing them to the mainland. This material is self-contained and thus appears first, preceding the chapters which draw on it.

We turn then to the distributions of single species and communities of species. This chapter is the least coherent, partly because it is easiest to proceed by listing examples; partly because the adaptations of species have large genetic and physiological components that we are not attempting to include.

Next we present a unified account of the various reasons one area has more species than another. Our effort is not to understand all cases of different numbers of species occupying different places. That would be too complex to be rewarding. Rather, our aim is to select certain cases that reveal interesting things about the mechanisms involved.

In tropics-temperate comparisons, we describe many, but by no means all, of the interesting contrasts between those very different parts of the earth. Some of these are unified by an understanding of competition while others remain mysterious.

Finally, in "The Role of History," we discuss some of the ways in which events of the past can be inferred. The aim is not to be exhaustive, which would take volumes, but rather to glimpse the methods. Events upon which history has left an indelible mark complete the chapter.

Island Patterns

5

In 1917 Howell published a summary of the birds known to breed on each of the Channel Islands off the coast of southern California (Fig. 5-1). About 50 years later Diamond (1969) conducted a new survey of the same islands. A comparison of Howell's and Diamond's results is most illuminating and provides our introduction to the patterns of island biogeography.

Diamond's table comparing the numbers of species in 1968 with those on the same island in 1917 is reproduced as Table 5-1. There are four remarkable features of Diamond's results.

First, no island has nearly the number of species it would have if it were part of the mainland. This is clear only to one who knows

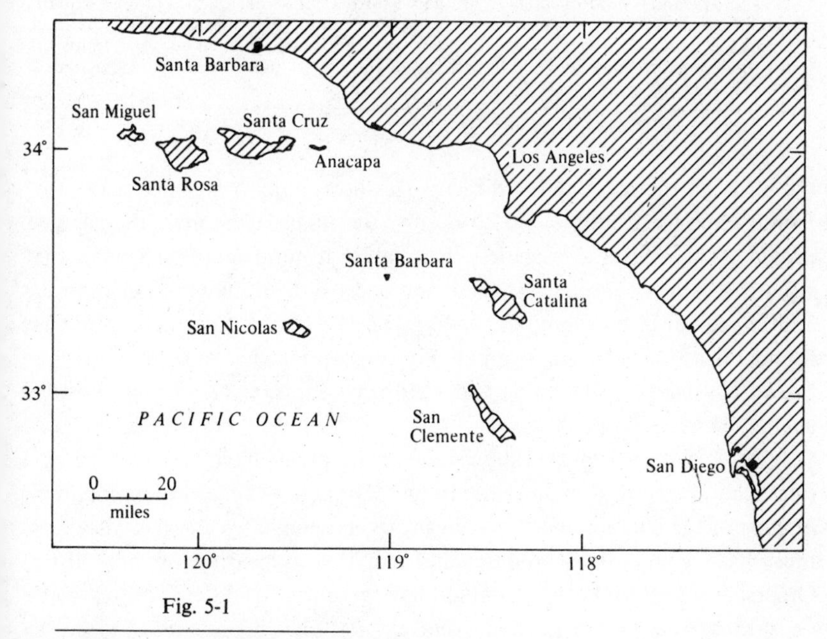

Fig. 5-1

Map of the Channel Islands off California.

Table 5-1
Avifaunal Turnover on the Channel Islands
(From Diamond, 1969)

	Area A	Dis-tance B	1917 species C	1968 species D	Extinc-tions E	Addi-tions F	Intro ductions G	Immi-grations H	Turn-over I
Los Coronados	1.0	8	11	11	4	4	0	4	36
San Nicolas	22	61	11	11	6	6	2	4	50
San Clemente	56	49	28	24	9	5	1	4	25
Santa Catalina	75	20	30	34	6	10	1	9	24
Santa Barbara	1.0	38	10	6	7	3	0	3	62
San Miguel	14	26	11	15	4	8	0	8	46
Santa Rosa	84	27	14	25	1	12	1	11	32
Santa Cruz	96	19	36	37	6	7	1	6	17
Anacapa	1.1	13	15	14	5	4	0	4	31

For each island, column A gives the area in square miles; B, the distance in miles from the nearest point on the mainland; C, the number of species of land and freshwater birds breeding in 1917; D, the number of breeding species in 1968; E, the number of species that were breeding in 1917 but not in 1968 and hence must have gone extinct in the interim; F, the number of species breeding in 1968 but not in 1917 ("additions"); G, the number of species present in 1968 that had been successfully introduced by man between 1917 and 1968 (all of these are game birds: California quail, Gambel's quail, pheasant, or chukar); H, the number of species present in 1968 but not in 1917 that had immigrated under their own power between 1917 and 1968, calculated as F minus G; and I, the turnover rate expressed in per cent of the species pool for 51 years, calculated as $100\,(E + H)/(C + D - G)$.

the birds on the mainland and recognizes how many more species (93 for Santa Cruz, for example) would occupy the range of habitats the islands provide. The islands average less than half that number. In particular, the wrentit (*Chamaea fasciata*) is the commonest bird in mainland chaparral, and chaparral is the dominant habitat on the islands. Yet the wrentit is found on no island. Other apparently appropriate yet missing birds are brown towhee (*Pipilo fuscus*) and California thrasher (*Toxostoma redivivum*).

Second, a large number of species have clearly gone extinct from the islands in 50 years. In fact, Diamond's record of extinctions is short of the true number since, in the intervening 50 years, many species must have gone extinct and recolonized, thus appearing on both lists. Others, absent from both lists, might have colonized and then gone extinct. Only those that were recorded, went extinct, and did not recolonize are included in Diamond's record of extinction. Yet there are a great many even of these. Roughly one-third (30%) of the bird species present on a given island in 1917 were missing from that island in 1968. A few of the extinctions can be assigned to obvious causes. Man's use of pesticides has eliminated the peregrine falcon (*Falco peregrinus*) from most of its range in the world, and the Channel Islands (like the rest of California) reflect this. But most of the extinctions, except on the island of Santa Barbara, show no obvious cause. (Santa Barbara Island was devastated by fires in 1959, and many habitats with corresponding birds were eliminated.) The species that went extinct from some island in the Channel group include Anna's hummingbird (*Calypte anna*), Allen's hummingbird (*Selasphorous sasin*), red-shafted flicker (*Colaptes cafer*), black phoebe (*Sayornis nigricans*), barn swallow (*Hirundo rustria*), red-breasted nuthatch (*Sitta canadensis*), raven (*Corvus corax*), burrowing owl (*Speotyto cunicularia*), white-throated swift (*Aeronautes saxatalis*), and lark sparrow (*Chondestes grammacus*). These are not species for which one can assign an easy man-made reason for extinction. Some of them immigrated onto other islands as they went extinct.

Third, Diamond's table shows an equally impressive number of new immigrants on each island. (As in his record of extinctions, his numbers are underestimates.) These are largely familiar mainland species and even species found on other islands. Again, although man is responsible for deliberate introductions of species like pheasants (*Phasianus colchicus*) onto the islands, most have no obvious explanation. (The islands of Santa Rosa and to a lesser extent San Miguel are exceptions, since the owner in 1917 was so "ornery" that Howell only got partial censuses by stealth. Many of those islands' "immigrants" are doubtless species present in 1917 but not then recorded.)

Fourth, and most remarkably, except for the island ravaged by fire and the ones not adequately censused in 1917, all the islands have very nearly the same number of species now as they did then. Thus the islands are not short of species compared to the mainland because the species never arrived; rather, new ones keep arriving and others keep going extinct, and the extinctions balance the immigrations, producing an equilibrium.

The same four properties of island species have been demonstrated in an equally remarkable experimental study carried out by Wilson and Simberloff (1969; Simberloff and Wilson, 1969, 1970). These authors chose four very small islands of red mangrove (*Rhizophora mangle*) in Florida Bay and had a professional exterminator erect a scaffold, enclose the island in a plastic sheet, and pump methyl bromide in until the animals were all killed, leaving the mangroves unharmed. Wilson and Simberloff had carefully recorded the species present on the islands before "defaunation," and after the exterminator had done his job the recolonizations were carefully monitored (Figs. 5-2 and 5-3). Remarkably enough, the islands acquired a fauna of the original number of species within about 6 months, but the names of many of the species were different from the

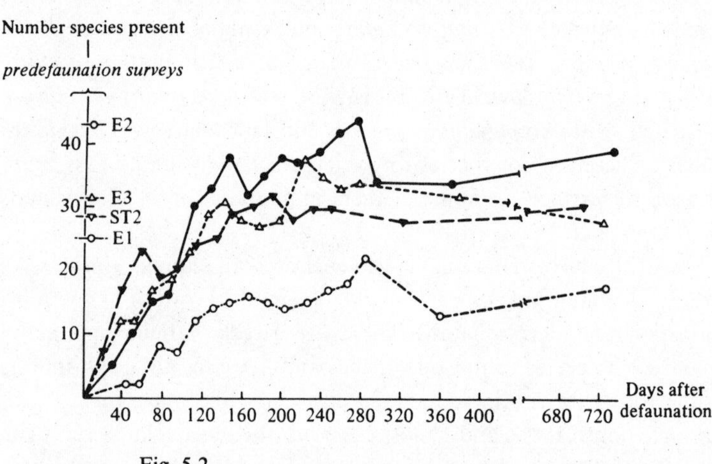

Fig. 5-2

The colonization curves of four small mangrove islands in the lower Florida Keys whose entire faunas, consisting mostly of arthropods, were removed by methyl bromide fumigation. The species numbers just before defaunation and at intervals following it are shown. The number of species is an inverse function of the distance from the nearest source. This effect was evident in the predefaunation census and was preserved when the faunas regained equilibrium after defaunation. Thus, the near island E2 has the most species, the distant island E1 the fewest, and the intermediate islands E3 and ST2 intermediate numbers of species. (From Simberloff and Wilson, 1970.)

Fig. 5-3

History of the colonization of E9, another typical experimental island in the Florida Keys. "Pre" is the predefaunation census. Solid entries indicate that a species was seen; shaded, that it was inferred to be present from other evidence; open, that it was not seen and was inferred to be absent. (From Simberloff and Wilson, 1969.)

The colonists of island E9

Days after defaunation

Pre | 24 | 45 | 62 | 84 | 101 | 117 | 136 | 153 | 171 | 193 | 210 | 229 | 247 | 266 | 364

Order	Family	Species
RTHOPTERA	Gryllidae	*Cycloptilum* sp.
		Cyrtoxipha confusa
		Orocharis sp.
ERMAPTERA	Labiduridae	*Labidura riparia*
OLEOPTERA	Anobiidae	*Cryptorama minutum*
		Tricorynus sp.
	Anthicidae	*Sapintus fulvipes*
		Vacusus vicinus
	Buprestidae	*Actenodes auronotata*
		Chrysobothris tranquebarica
	Cantharidae	*Chauliognathus marginatus*
	Cerambycidae	*Styloleptus biustus*
	Curculionidae	*Cryptorhynchus minutissimus*
		Pseudoacalles sp.
	Lathridiidae	*Holoparamecus* sp.
	Oedemeridae	*Oxacis* sp.
	Fam. Unk.	Gen. sp.
HYSANOPTERA	Phlaeothripidae	*Haplothrips flavipes*
		Neurothrips magnafemoralis
	Thripidae	*Pseudothrips inequalis*
ORRODENTIA	Caeciliidae	*Caecilius* sp. np.
	Lachesillidae	*Lachesilla* n. sp.
	Lepidopsocidae	*Echmepteryx hageni b*
	Liposcelidae	*Belaphotroctes okalensis*
		Embidopsocus laticeps
		Liposcelis sp. not *bostrychophilus*
	Peripsocidae	*Ectopsocus* sp. *bμ*
		Peripsocus stagnivagus
	Psocidae	*Psocidus texanus*
		Psocidus sp. 1
	Trogiomorpha	Gen. sp.
EMIPTERA	Anthocoridae	*Dufouriellus afer*
	Cixiidae	*Oliarus* sp.
	Miridae	*Psallus conspurcatus*
	Pentatomidae	*Oebalus pugnax*
	Fam. Unk.	Gen. sp.
EUROPTERA	Chrysopidae	*Chrysopa collaris*
		Chrysopa externa
		Chrysopa rufilabris
EPIDOPTERA	Eucleidae	*Alarodia slossoniae*
	Olethreutidae	*Ecdytolopha* sp.
	Phycitidae	*Bema ydda*
	Psychidae	*Oiketicus abbottii*
	Ptineidae	*Nemapogon* sp.
	Pyralidae	*Thoieria reversalis*
	Saturniidae	*Automeris io*
	Fam. Unk.	Gen sp.
IPTERA	Hippoboscidae	*Olfersia sordida*
	Fam. Unk.	Gen. sp.
YMENOPTERA	Braconidae	*Apanteles hemileucae*
		Apanteles marginiventris
		Callihormius bifasciatus
		Ecphylus n. sp. nr. *chramesi*
		Iphiaulax epicus
	Chalcidae	Gen. sp. 1
		Gen. sp. 2
		Gen. sp. 3
		Gen. sp. 4
	Eulophidae	*Euderus* sp.
	Eumenidae	*Pachodynerus nasidens*
	Eupelmidae	Gen. sp.
	Formicidae	*Brachymyrmex* sp.
		Camponotus floridanus
		Camponotus sp.
		Crematogaster ashmeadi
		Monomorium floricola
		Paracryptocerus varians
		Pseudomyrmex elongatus
		Pseudomyrmex "flavidula"
		Tapinoma littorale
		Xenomyrmex floridanus
		Gen. sp.
	Ichneumonidae	*Calliephialtes ferrugineus*
		Casinaria texana
	Pteromalidae	*Urolepis rufipes*
	Sphecidae	*Trypoxylon collinum*
	Vespidae	*Polistes* sp.
RANEAE	Araneidae	*Argiope argentata*
		Eriophora sp.
		Eustala sp.
		Gasteracantha ellipsoides
		Metepeira labyrinthea
		Nephilia clavipes
	Clubionidae	*Aysha* sp.
	Dictynidae	*Dictyna* sp.
	Gnaphosidae	*Sergiolus* sp.
	Linyphiidae	*Meioneta* sp.
	Lycosidae	*Pirata* sp.
	Salticidae	*Hentzia palmarum*
	Scytodidae	*Scytodes* sp.

original ones. After the first 6 to 9 months the number of species no longer increased (perhaps there was even an overshoot followed by a significant decrease) but new species continued to replace old ones. These experiments not only confirm the four points we drew from Diamond's study, but add an additional one.

Fifth, the departure of the number of species present from the original number decreased about exponentially at such a large rate that 6 months sufficed to achieve virtual equilibrium. If the overshoot was real, it probably reflects the fact that all the species were initially rare; when their abundances grew, their interactions increased and some were eliminated by competition or predation, causing the decrease.

In the following sections of this chapter we investigate these five, as well as other, properties of islands in greater detail. The book by MacArthur and Wilson (1967) treats many of these subjects at considerably greater length, although it was written before many of the researches we discuss here were conducted.

The Nature of Island Colonization

There is a large element of chance in successful island colonization. The growth of the founding population must be very smooth, any sharp dips in a very small population being fatal. This requires good luck, since small ups and downs are normal during population growth. The actual arrival of the colonists also requires good luck. Flying birds and insects may be able to direct their passage to the island, but many other organisms have more passive dispersal. Faced with the improbability of mammals or lizards or even of sedentary birds reaching islands, biogeographers used to postulate former land bridges connecting the mainland to the island. Although many islands were once connected to the mainland and some still have species remaining from the days of the connection, it is astonishing what can be accomplished without a land bridge.

The Smithsonian Institution Center for Short-Lived Phenomena (reports from various observers: Numbers 662, 664, 665, 666, 671, 681) reported the history of a remarkable floating island that apparently broke loose from eastern Cuba on July 1 or 2 of 1969. It was first seen 60 miles south of the Guantanamo Naval Base, and by July 11 it had moved to latitude 19°34′N, longitude 75°14′W, about one-quarter of the distance to eastern Jamaica. This island was estimated to be 6–12 m tall and 13 m in diameter, containing 10 to 15 trees, their bases tangled with lower vegetation. By July 19 an aerial search failed to locate any part of the island, and a 12-m-long, 60-cm-in-diameter palm tree floating upright with 3 m of its length above water on July 15 was probably a remnant of it. Unfortunately, no one saw the island at close range and no one knows whether lizards, birds, or even mammals were hidden in its vegetation, but this occurrence does demonstrate how very heavy animals incapable of swimming could undoubtedly be transported large distances across open ocean. This is not to claim that rhinoceroses floated out to Java; we know Java was connected to the mainland in recent geologic times, and the rhinos doubtless walked. But it does mean that merely because an island has some sedentary mammal or bird, biogeographers should not overrule geological evidence that this island was unconnected.

The Pearl Islands (Archipielago de las Perlas) south of Panama about 50 miles are on the continental shelf (Fig. 5-4). In fact, an ocean fall of about 37 m would connect all of its islands together and to the mainland. Since the ocean level was about 100 m lower than it is now during the recent glaciation, when much water was bound up in glacial ice, we conclude that the Pearl Islands were recently connected to the Panama mainland. In fact, when the islands were connected, man was already in Panama and could walk out onto what are now the islands. Although they were connected so recently and are largely uninhabited by man, and largely covered with heavy forest, their bird species are astonishingly like those on the truly oceanic West Indies, which were never connected to the mainland. Not only have the Pearl Islands lost most of the mainland species they must have had at the time of the connection, but they have apparently been recolonized by the good colonist species that independently proved their colonizing ability in the West Indies. We return later to other aspects of the Pearl Island species, but here we note what kind of species proved successful, following MacArthur, Diamond, and Karr (in press). Karr (1971) classified the resident birds of mainland Panama into several habitat types including "early shrub" (a bushy field), "late shrub" (later in the succession when some trees were tall but the canopy was not closed), and "moist forest" (a closed canopy and tall trees). The 70-hectare island of Puercos in the Pearl Archipelago had only 20 species of birds although it was covered with heavy forest (just less dense and tall than the "moist forest" of the mainland). Yet, every species is from low second growth on the mainland. For instance, mainland forests like those on Puercos would have the slaty antshrike (*Thamnophilus punctatus*) whereas Puercos had, and was literally crawling with, the barred antshrike (*T. doliatus*), which is confined on the mainland to second growth (Fig. 5-5). The commonest flycatcher on the island was the scrub flycatcher (*Sublegatus arenarum*), which is confined, as its name suggests, to scrubby habitats on the mainland. Now, why are second growth species such good colonists and such

Fig. 5-4

Map of the Pearl Islands, Panama.

Fig. 5-5

Barred antshrike (*Thamnophilus doliatus*). This mainland second growth bird has proved itself to be a very good colonist. (After Wetmore, 1957.)

good persisters after they colonize? We can only guess that these species are adapted to changing habitats; second growth disappears at one place as it turns into forest and reappears at another place where a field is abandoned. The second growth species, then, must frequently move to a new location and must tolerate a moderate variety of habitats, while a species of deep forest need never move and can specialize to a very particular habitat. For these reasons it would indeed seem plausible that the second growth species should be good and successful colonists.

Much earlier, in fact, Wilson (1959, 1961) had classified the ants he knows so well into those that are very good colonists, as evidenced by their expansion, continuous ranges, and small geographic variation, and those that are slower colonists. He called the expanding species "stage I" species and found that they occupy marginal habitats on the mainland, specifically open lowland forest, savannah, monsoon forest, and littoral zone (Fig. 5-6). Wilson also showed (Fig. 5-7) that they occupy

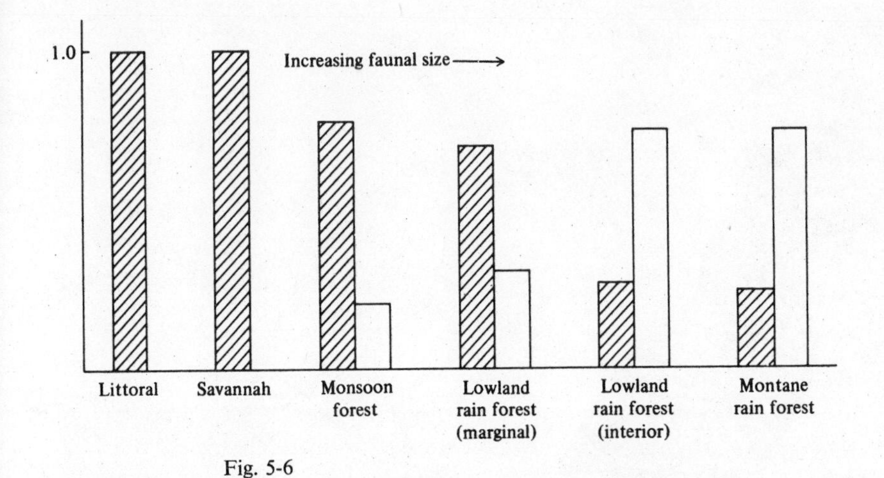

Fig. 5-6

Proportion of stage I ant species (good colonists), shaded bars, and of other ant species, open bars, in different mainland New Guinea habitats. (After Wilson, 1959.)

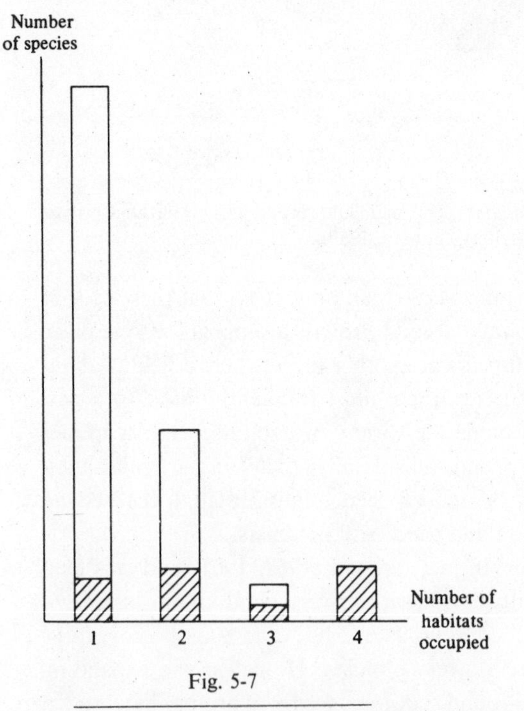

Fig. 5-7

Stage I ant species, shaded bars, and other ant species, open bars, showing that stage I species occupy more habitats than other species do on the mainland of New Guinea. (After Wilson, 1959.)

a wider range of habitats on the mainland than other ant species and are in this sense, too, broader. Finally, Wilson studied these stage I species on the small island of Espiritu Santo, where, in the absence of competitors, they became the dominant species even in deep virgin rain forest. Thus they expand their habitats when released from competition. These results exactly parallel the bird observations from Puercos.

To add yet another example, we note that E. E. Williams (1969), who, with his students at Harvard, has been carrying on fine and detailed studies on West Indian lizards, noted the same thing for *Anolis carolinensis*, the green lizard that has colonized so successfully in the West Indies. This lizard, like the successful ants and birds, is tolerant of many conditions and is not a deep forest species.

Summary of Colonization Theory

Theory gives additional clues to the nature of successful colonization. To grow past the vulnerable stage rapidly, a founder population should have as large as possible a rate of increase in population. The rate of increase, r, is the difference between per capita birth rate, λ, and per capita death rate, μ. But notice that the same rate of increase, r, can be achieved by large birth rate and large death rate just r less, or by small birth rate and very small death rate, again just r less. Of these alternatives, the latter is clearly safer.

Every bounded population is sure to go extinct eventually. But populations that cannot exceed some very small size, K, are likely to go extinct much faster than those that can reach very large populations, K. Figure 5-8 shows, in terms of λ, μ, and K, the expected time it would take a population founded by a single pair to go extinct. The derivation of these results is given in the chapter appendix, together with the population growth assumptions used. Other assumptions might lead to slightly altered curves, but the general picture would doubtless remain essentially the same. Basically, the figure shows that for small K (i.e., K values in tens or small hundreds) extinction is likely to be fairly rapid, whereas for K values in thousands or more, the expected time to extinction is very large

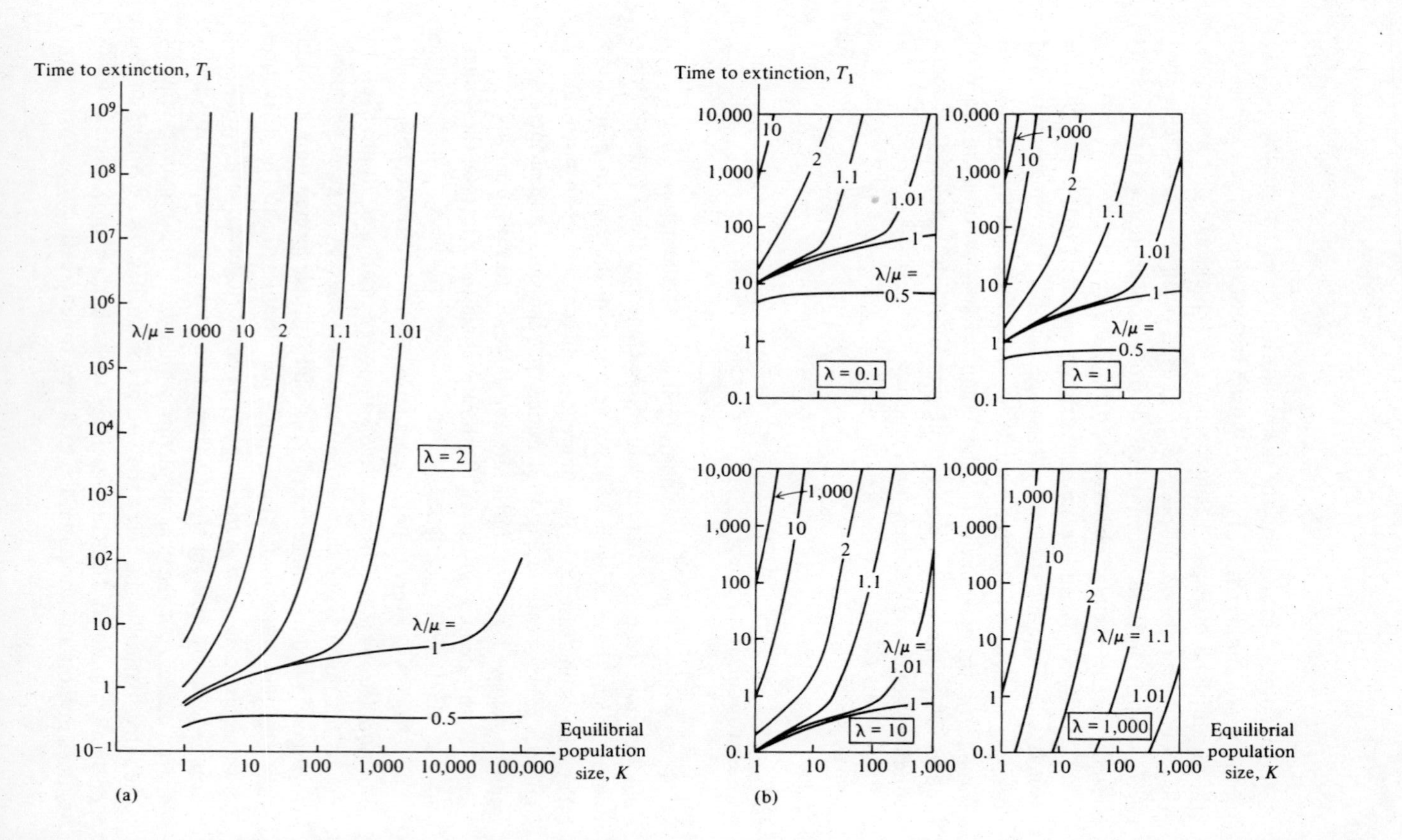

Time to extinction, T_1
λ/μ = 1000
10
2
1.1
1.01
λ = 2
λ/μ =
1
0.5
Equilibrial population size, K
(a)
Time to extinction, T_1
λ = 0.1
λ = 1
λ = 10
λ = 1,000
λ/μ =
0.5
1.01
λ/μ = 1.1
Equilibrial population size, K
(b)

indeed. It is so large in fact that populations with K's in the hundreds of thousands can usually expect to last as long as the age of the universe. Such populations are "safe" while those with smaller K's are "vulnerable," and the switch from vulnerable to safe occurs quite rapidly as K's increase. This explains why populations on small islands (small K's) are so likely to go extinct and those with large K's, remaining on large land bridge islands, persist so long.

In the appendix we also derive a result that shows how dramatic is the effect of a competitor on a species' survival time. Consider a competitor which, on the mainland, would reduce a species population size to two-thirds of what it would be without the competitor. This is not very severe competition, yet note its effects if the competitor is established first on an island. The expected survival time of a new colony of a species on an island is reduced by the competitor to roughly the cube root of what it would be without competition. If a pair of a species and its descendants expect to last 1000 years on an island without the competitor species, they would expect to last only 10 years on a similar island on which the competitor was established.

This gives us a new "island view of competition," developed further in the final section of the chapter, to supplement the view of competition developed in Chapter 2. The competition equations (Chapter 2 appendix) describe the very gradual and deterministic replacement of one species by a superior competitor. In the island view, the populations are fragmented, are beset by frequent local extinctions, and find it very difficult to recolonize where an effective competitor is present. Hence the replacement of one species by another may be rapid and accompanied by abrupt local extinctions.

Fig. 5-8

(a) T_1, the expected (mean) survival time of a population beginning with a single propagule, as a function of the per capita birth rate λ, per capita death rate μ, and maximum number K of individuals belonging to the species that the island can hold, where $\lambda = 2$. The propagule is defined as the minimal number of individuals capable of reproducing—ordinarily a gravid female, a seed, or an unmated female plus male. These curves are based on the condition of density-dependent death. (b) Families of curves similar to that in (a) but with differing values of the per capita birth rate λ. (From MacArthur and Wilson, 1967.)

The Nature of Extinction

Some aspects of extinction were included in our discussion of colonization, since a colonist isn't successful unless it avoids extinction. But there are additional aspects that must be considered. We have witnessed hundreds of species' extinctions during human history. Can we assume these are typical of all extinctions and, in particular, of all island extinctions? And as a matter of general interest, do we know what caused large-scale extinction in the past? Why, for instance, did most of the large grazing species—camels, elephant relatives, giant sloths, etc.—go extinct about 10,000–12,000 years ago in North America? (See Fig. 5-9.)

We first consider what competition can do. We concluded in Chapter 2, in which we discussed the machinery of competition, that the closer two species are and hence the more they compete, the smaller the geographical zone of overlap between them. But unless one species is uniformly superior to another in every habitat, it will not exterminate the other completely. That is, the usual effect of introducing a competitor is to restrict but not eliminate the previous species. On islands, especially small islands, even a restriction is likely to prove fatal. Suppose prior to competition species A lived in both forest and second growth but when competitor B was added, A was restricted to second growth with B occupying the forest. On the mainland nothing more would happen; with the introduction of B, A would contract but both would persist somewhere. On the island, only forest may be present, with no second growth refuge for species A. Then the introduction of B would exterminate A from the island. Even if there were a small patch of second growth, the population of A, when restricted to this habitat, would be so low that it would be vulnerable to any kind of accident, and extinction would be likely to ensue. In either case, competition that on the mainland would only restrict the habitat of a species may, on an island, cause its extinction.

What, then, traditionally causes mainland extinction? Predation, or its equivalent, disease, a form of predation, certainly can cause outright extinction. Suppose r is the largest value the relative rate of increase of a population can take. This is the "interest rate" at which a population grows. Just as an imprudent investor who withdraws from his savings faster than interest accumulates will lose his savings, so a popula-

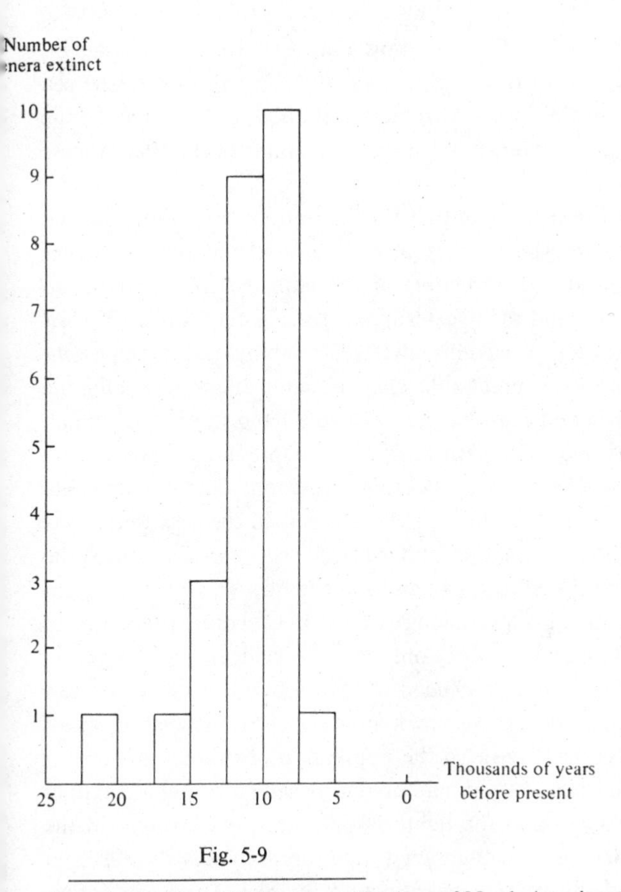

Fig. 5-9

Chronology of latest Pleistocene extinctions of North American land mammal genera. (From Webb, 1969, after Hester.)

tion harvested at a rate greater than r by predators will go extinct. Although we know astonishingly little about the precise causes of extinction of the species we have seen disappear, predation, including disease, surely plays its part. The sea otter (*Enhydra lutris*), thought for years to be extinct, was reduced solely by the great pressures man as a predator exerted on it (mostly for its fur). Once the sea otter was thought extinct, it was easy to induce Congress to pass a bill protecting it, for no economic interests then opposed the bill. Fortunately, a few sea otters had remained, and with their new protection they have increased to the point where one may go to see them in California waters. Of course, now that they are common again, they have raised the wrath of the abalone fishermen, who claim the sea otter's taste for abalone interferes with their livelihood, and again economic interests are at work. However, this is a classic example of the effectiveness of predation.

At the same time that the fur traders discovered the rich fur of the sea otters in Alaska, they also discovered the curious Steller sea cow, a dugong and larger relative of the manatee still surviving in southern estuaries. Man did really exterminate these, either for food or for fun. Man the predator has proved his effectiveness many times; the passenger pigeons and heath hens, great auks and Labrador ducks, have all gone extinct in recent times and man has had a strong hand in each, although other events have played their part. And man the predator surely helped to reduce the bison of the prairies to virtual extinction. This human effectiveness as a predator has led Martin (1971) to defend the view that it was the arrival of man in North America about 11,000 years ago that caused the extinctions of mammoths, sloths, camels, and the like. In any event, man is a good predator, capable of producing extinction. Can other predators do the same? On the mainland, possibly only rarely. The degree of synchrony and orderliness of the predation needed to cause complete extinction can probably only be regularly achieved by man. Even the famous myxomatosis virus, which enormously reduced the population of rabbits in Europe, left many. But on an island, nonhuman predators can and do cause extinctions. On tiny Stephen Island the lighthouse keeper's cat brought in the only known specimens of the Stephen Island wren (*Xenicus lyalli*) and doubtless singlehandedly caused its extinction in 1894. Feral cats, mongooses, and Norway rats have unquestionably all caused island extinctions. The island populations are not only more vulnerable because they are smaller and confined to fewer habitats, but also because they have often evolved in the absence of predators. When man or his accomplices in the form of cats, rats, or mongooses arrives, the native fauna is totally unprepared. Figure 5-10 shows how both man as a predator (hunting eggs and feathers) and larger gulls that compete for nest sites and also eat laughing gulls have twice nearly exterminated the laughing gulls from Muskeget Island, Massachusetts.

Even when predation does not cause extinction, it may have dramatic effects on population levels of the prey. This is seen most

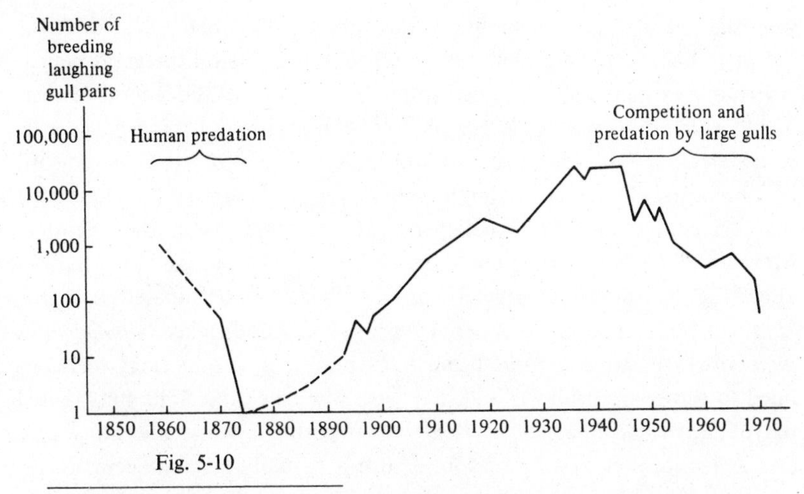

Fig. 5-10

Numbers of breeding laughing gull (*Larus atricilla*) pairs estimated to be present on Muskeget Island, Massachusetts, in various years. Before 1870 there were "many" and between 1874 and 1893 the populations were of unknown size; these are shown by dashed lines. Recent increases in herring gulls (*L. argentatus*) and great black-backed gulls (*L. marinus*) are presumably associated with man's increased production of garbage. The increase of these large gulls is a major factor in the current decline of the laughing gull. (After data of Nisbet, 1971.)

strikingly in insects and other pests introduced without their natural predators. Elton's (1958) book is full of examples. One of the most spectacular involved the prickly pear cactus (genus *Opuntia*) introduced into Australia without natural predators. It soon overran Australia, causing almost unbelievable thickets. When the moth *Cactoblastis cactorum* was introduced, it reduced the cacti to a manageable level.

We have now seen three causes of extinction: random population fluctuations, competition, and predation (including disease). The random extinctions are likely and important only when the populations are low, but both competition and predation can reduce populations enough so that random causes can take over. There is, however, one more major cause of population reduction that can cause direct extinction and can certainly help to lower the population to the precarious level. This is habitat alteration. The early stages of the heath hen's extinction were due to predation and habitat alteration, although the extinction appeared to be completed finally by random events. Many species are confined to a fairly well-marked habitat, including food supply. The black-footed ferret (*Mustela nigripes*), whose habitat is prairie dog towns on the great plains, is virtually extinct. This is partly through the courtesy of the federal government, which does wholesale poisoning of the prairie dogs and hence kills some ferrets, but it is more the work of man the predator, who (via poison) is eliminating the ferrets' habitat; prairie dog towns are so far apart now that ferrets may not be able to disperse between them. As another example, Mexicans, like Americans, cut their virgin pine forests, and this habitat alteration has either completely or very nearly exterminated two fine Mexican birds. The great imperial ivory-billed woodpecker (*Campephilus imperialis*), the world's largest and most powerful woodpecker, was confined to these virgin forests and may well now be extinct. The thick-billed parrot (*Rhynchopsitta pachyrhyncha*), which formerly wandered into southeast Arizona after nesting season, is also a casualty since it seems to require woodpecker holes for its nests.

We now give a sample of the kind of extinction problem that plagues biogeographers. We showed that Puercos Island and the Pearl Archipelago in general, south of Panama, were connected to the mainland within 10,000 or 12,000 years ago and must then have had most of the appropriate mainland bird species. Yet by now 19 complete families of mainland birds have gone extinct from the islands. Prominent among these are trogons, with 11 mainland Panama species, puffbirds with 8 mainland species, toucans with 6 mainland species, woodcreepers with 17 species on the mainland, ovenbirds with 22 species on the mainland, cotingas with 20 species on the mainland, manakins with 12 species on the mainland (including the commonest species in Karr's plots and hence a likely candidate for Puercos), and thrushes with 11 species on the mainland. Many of the species of these families must have been present on the Pearl Islands within 12,000 years. Why have they all gone extinct? Diamond (in press) has shown that on two oceanic island systems the species that have gone extinct are usually those with low abundances (small K), or nearby competitors (large α), in accord with the theory of random extinction discussed on pages 89–91 and in the chapter appendix.

So far we have viewed islands as places where extinctions are greatly accelerated; most of the species man has seen exterminated are island species. But there is another side to the coin. Species on islands are often astonishingly abundant, being free from predators and often from competitors. The West Indies are literally teeming with lizards of the genus *Anolis* and also, for example, with the flower feeding bird called the "bananaquit" (*Coereba flaveola*). Both *Anolis* and *Coereba* are much rarer on the Central American mainland. We might guess that if some new hazard to the life of these species appeared, it would perhaps be on the islands that they would survive. In any case, there are some species that do survive solely on islands although they were once more widespread. Thus there are *Solenodon* (large insectivorous mammals) remaining only on Cuba and Hispaniola. The fossil remains of related forms are found on the mainland, showing that they were once more widespread. For that matter, the American ivory-billed woodpecker (*Campephilus principalis*), like its Mexican cousin, is nearly extinct and may make its last stand in the pine forests of Cuba, alongside *Solenodon*. Of course, there are vast numbers of species that originated on islands; it would not be fair to use their present distributions as evidence of long life of some island species. Dodos and moas cannot be said to have survived longer on islands, because they never occurred on the mainland. But several species of birds that were formerly widespread on New Zealand are now confined to its tiny offshore islands. Examples are the flower feeding stitchbird (*Notiomystis cincta*) and the omnivorous saddleback (*Philesturnus carunculatus*). Man has introduced many species of predators to New Zealand and also many potentially competing bird species. The two species mentioned doubtless thrive on the offshore islands in part because predators are scarce or absent.

Island Equilibrium

The studies of Diamond and of Wilson and Simberloff show that the number of species on an island reaches an equilibrium where extinction balances immigration. Such an equilibrium is no accident—rather we expect each island to be approaching some sort of equilibrium. The reason each island has an equilibrium is shown in Fig. 5-11. As the number of species present on the island increases, fewer of the immigrants are new, so the immigration rate of new species falls, reaching zero when all mainland species are present. Furthermore, as the number of species increases, the rate of extinction increases for two reasons. First, there is

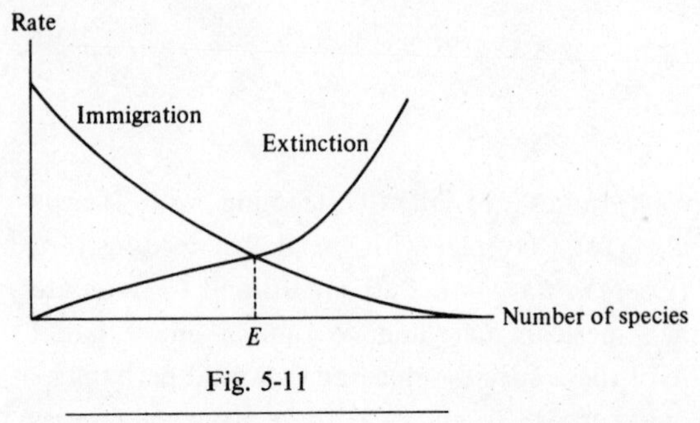

Fig. 5-11

The number of new immigrant species falls and the number of extinctions rises as the number of species on an island increases. When the island has E species, these two processes balance and the island is at equilibrium. The sharp rise in the extinction curve occurs when all island habitats are occupied by appropriate species. See the text for details.

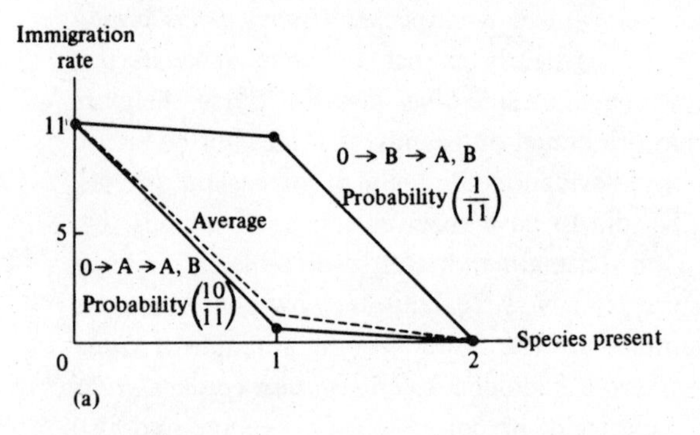

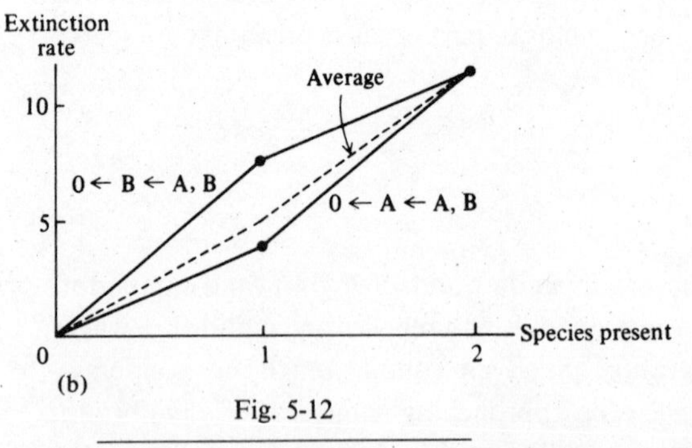

Fig. 5-12

(a) Immigration curves for the two orders of arrival of two species, A with immigration rate 10 and B with immigration rate 1. The average curve is $\frac{10}{11}$ times the $0 \rightarrow A \rightarrow AB$ immigration curve plus $\frac{1}{11}$ times the $0 \rightarrow B \rightarrow AB$ curve. Notice that, as in Fig. 5-11, the average curve is steeper on the left. (b) Extinction curves for the two orders of extinction $0 \leftarrow A \leftarrow AB$ (B goes extinct before A) and $0 \leftarrow B \leftarrow AB$. We suppose the extinction rate of B is 8 and that of A is 4 on the same scale. The average is calculated as in (a).

an increasing number of species to go extinct, and second, each species may be rarer (because of competition) when more species are present, so each may have an increased chance of extinction. In any case, the extinction rate rises from zero, when there are no species to go extinct, to a higher level. Now, a curve falling to zero must intersect another rising from zero, and at the intersection, immigration and extinction are equal. If the number of species is larger than equilibrium, extinction exceeds immigration, causing a decline; if the number of species is less than equilibrium, immigration exceeds extinction and the number of species increases. Hence this is a stable equilibrium. This view of equilibrium is useful and gives insight into various island patterns, but it is also misleading in one respect that we must clear up.

Suppose, to make the matter as clear as possible, that there are only two mainland species available for colonizing the island. Species A has an enormous immigration rate, and species B a very low one; to be precise, we assume A has immigration rate 10 and B, 1. (The immigration rate can be interpreted as the number of islands out of 100 identical islands on which immigrants of the species would arrive each year.) Now there can only be 0, 1, or 2 species on the island. Zero is unambiguous and so is 2 (both A and B must be present), but when only 1 species is present, it can either be A or B. And when the 1 species is A, the rate of immigration of new species is the rate of immigration of B (i.e., 1); while if the 1 species is B, the rate of immigration is the very high rate, 10, of arrival of A. In other words, there are two immigration curves, and they coincide for 0 and 2 species but are different for 1. There are two immigration curves because there are two orders of arrival: first A then B, and first B then A (Fig. 5-12). More generally, if there are 10 species available, there will be as many immigration curves as there are orders of arrival of the immigrant species. The first arrival can be any of the 10; then for each of these possibilities the second can be any of the remaining 9, and so on. Hence there would be $10 \cdot 9 \cdot 8 \cdot 7 \cdot 6 \cdot 5 \cdot 4 \cdot 3 \cdot 2 \cdot 1 = 10!$ (read "10 factorial") different immigration curves. If there are N species, there are $N!$ different immigration curves. Each of the immigration curves has a probability: In our first example A was 10 times as likely to be the first immigrant as B, so the order "A then B" has probability $\frac{10}{11}$ and the order "B then A" has the remaining probability $\frac{1}{11}$. The lower curves in the figure are more probable. We could also, although it would be more complicated, assign probabilities to each of the $N!$ immigration curves for N species. The immigration curve of Fig. 5-11 can be viewed as the average of all the separate curves for different orders of arrival—the average weighted according to their probabilities. For exactly analogous reasons there are $N!$ orders of extinction and hence $N!$ extinction curves, each with its probability, and the extinction curve of Fig. 5-11 is their weighted average. To make the problem clearer, an island might reach equilibrium along one immigration curve; suppose then that there are two extinctions and not of the last two species to arrive. In that case the recolonization will be in a different order of arrival and along a different immigration curve. Hence a new equilibrium could be reached.

We can either view Fig. 5-11 as the picture of what we expect but allow some variation, or more precisely we could superimpose all the extinction and immigration curves. These would come in pairs: there would be an immigration curve for order of arrival A, B, C, D, E, . . . , Z and a corresponding extinction curve which, beginning with all species, proceeded leftward from Z back to A. For each order of arrival there is an order of extinction that removes them in exactly the reverse order. The intersections of all of these pairs of curves are possible equilibrium points for the island. The island can actually hold at equilibrium any number of species between the leftmost and rightmost of the equilibrium points. The lesson of this digression is that the immigration and extinction curves of Fig. 5-11 should be viewed not as single sharp curves but as broad blurs, forming a reasonably large area of intersection. Bearing this in mind, we return to the simple things that can be learned from them.

First, we compare the equilibrium number of species on near and far islands that are otherwise similar (Fig. 5-13). Clearly, the far island has a lower immigration curve and has fewer species at equilibrium. Hence even at equilibrium remote islands will contain fewer species. Of course, remote islands with their lower immigration rates also take longer to reach equilibrium. There is no reason to assume that all islands have reached equilibrium. We have seen that the Channel Islands have reached it for birds and that Wilson and Simberloff's bits of mangrove islet reached equilibrium in a matter of months, not years. But who is to say that the rate of extinction of, say, mammals from some remote island is automatically equal to the rate of immigration? While some nearby islands are doubtless at equilibrium, especially with highly mobile forms

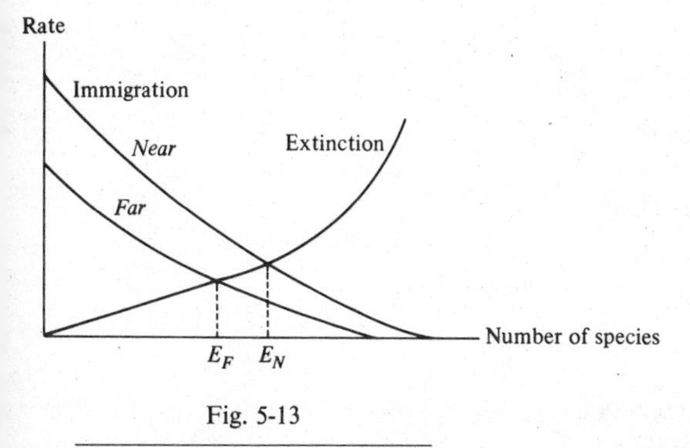

Fig. 5-13

A far island equilibrates with fewer species, E_F, than an otherwise identical near one, E_N.

like birds and flying insects, remote islands must often still be approaching equilibrium. We see later (pp. 107ff.) that many islands on the continental shelf may have more than the equilibrium number. We already have two reasons to expect farther islands to have fewer species: They would equilibrate with fewer, and they are less close to equilibrium. There is a sort of feedback that accentuates this. If the remote island has fewer plants, for either reason, it is not equivalent to a nearer one from an insect's point of view. Fewer kinds of insects could be supported there. This further reduces the remote island equilibrium, adding a third reason for remote islands to have fewer species: they may be more impoverished in their habitat.

Next, we turn to two islands the same distance from the mainland and not using one another as stepping stones for colonization. The only difference between the two is that we assume one island to be much larger (Fig. 5-14). The larger island may have a slightly higher immigration rate because it forms a larger target for immigrants to hit, and it will almost surely have a lower extinction rate for all the reasons we considered when we discussed extinction. Hence, as we show in the figure, the larger island will hold more species at equilibrium. Of course, we must remember that these are blurry lines; it would be possible for the large island actually to hold fewer species, although this would probably be quite temporary unless the island were more impoverished in its habitat, due, say, to its geology. In this case, its extinction rate would go way up and we would

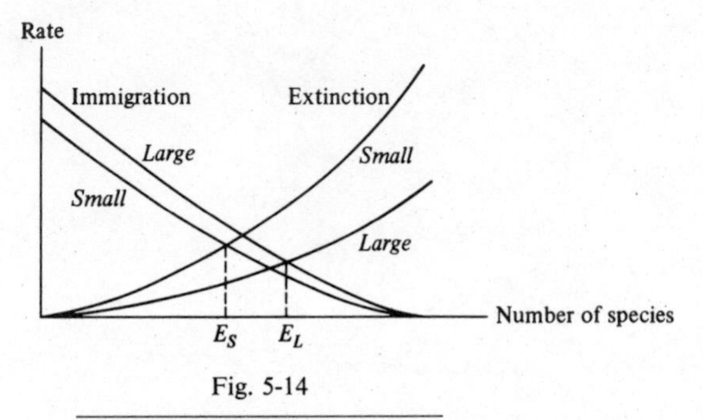

Fig. 5-14

A large island equilibrates with more species, E_L, than a small island, E_S, the same distance from the mainland.

understand its species impoverishment. Increased complexity, like increased size, may decrease the extinction rate and increase the number of species present.

There are many other inferences we can draw from such graphs: the effects of stepping stones, the comparative effect of area on remote compared to near archipelagos, faunal similarity of islands, etc. (MacArthur and Wilson, 1967). We will not dwell on these but will demonstrate the expected rate of approach to equilibrium. As shown in Fig. 5-15, if the immigration and extinction curves are straight, then the

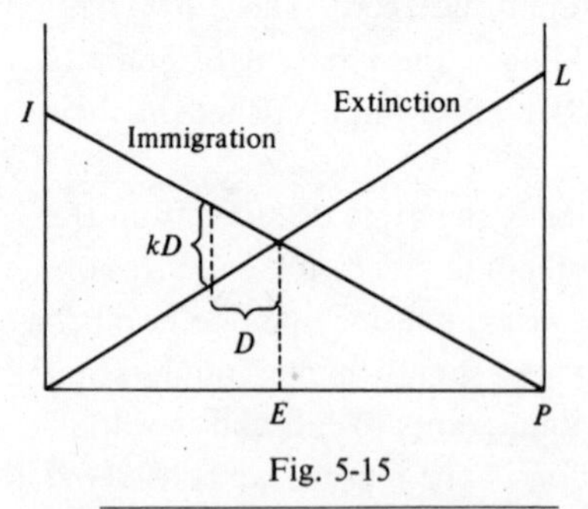

Fig. 5-15

I (or immigration rate when no species are present), P (or the species pool of available immigrants), and L (or the extinction rate when P species are present) are related in the figure. When the number of species departs by distance D from the equilibrium, E, the rate of increase in species—which is the negative rate of increase of D—is given by kD. In fact $k = \frac{I}{E} = \frac{L}{P - E}$. We can even solve this for $E = \frac{IP}{L + I}$, but for cases where the curves are not straight, the result does not hold.

difference between immigration and extinction rates—and this difference measures the net accumulation of species—is proportional to the departure, D, from equilibrium. We thus set this rate of accumulation of species equal to kD. In symbols, since the growth of the number of species is the negative of the change in the departure, D, $\frac{dD}{dt} = -kD$, whence D is of the form e^{-kt}, showing the exponential approach to equilibrium demonstrated experimentally by Wilson and Simberloff. If the curves were not straight this would not be strictly true.

Wilson and Taylor (1967) have added an evolutionary twist to the equilibrium model (Fig. 5-16). The first species to arrive and equilibrate are the "tramp" species, good immigrators, but inappropriate

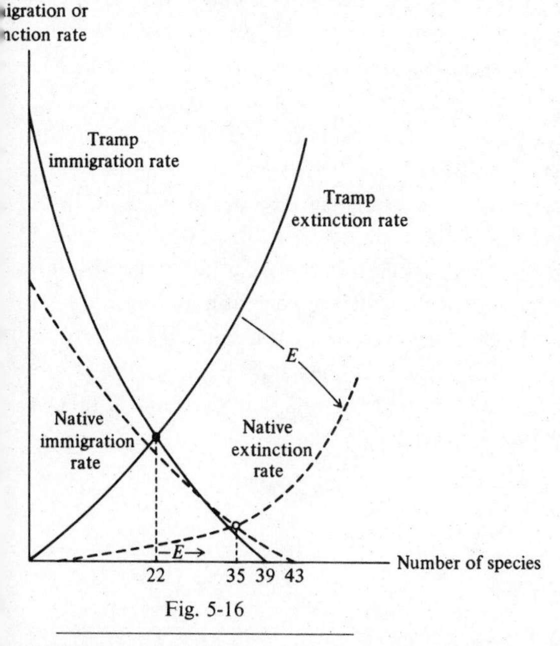

Fig. 5-16

Equilibrium model of the ant fauna of Upolu, Samoa. An attempt is made to estimate the probable increase in the number of species of tramp origin from the present interactive (or assortive) equilibrium, at 22 species, that would result if the extinction rate were lowered by means of evolution ($E\rightarrow$) toward the native level. The number of known native species on Upolu is 35, and this number is taken as the minimum to which the tramp equilibrium could move if given enough time. (From Wilson, 1969, after Wilson and Taylor.)

to the vegetation on the island. But this equilibrium is vulnerable to gradual replacement by more specialized native species. Hence there is, during the colonization of an island, an early "quasi-equilibrium" that progressively grows to a final equilibrium of the rich native fauna. Wilson and Taylor show, by way of example, that Polynesian islands with only tramp species of ants equilibrate—the early quasi-equilibrium—with very few species, while an island that has been colonized by a large, well-adapted native ant fauna has nearly double the number of species.

Thus the equilibrium theory embodies the five main results of the studies of Diamond and of Wilson and Simberloff.

The actual number of species on islands is often well predicted from area alone (Fig. 5-17), but on other occasions it is more closely related to the diversity of habitats. To incorporate this into equilibrium theory, we note that the extinction curve begins to rise abruptly only when there are species for all habitats and new species are homeless and have high extinctions. Thus complex habitats have lower extinction curves. Fairly detailed empirical studies of the effects of area, elevation, and distance on the number of species have been carried out by Hamilton, Barth, and Rubinoff (1964), who used the statistical technique of multiple regressions to estimate the roles of area, island elevation, and island isolation in determining the numbers of bird species. In the West Indies, for instance, area alone seemed to account for 93% of the variation among islands in numbers of bird species. In the East Indies, area accounted for 72% and elevation for an additional 15%. Such studies exhibit the easiest way to estimate the relative effects of different factors, but the models take no account of immigration and extinction.

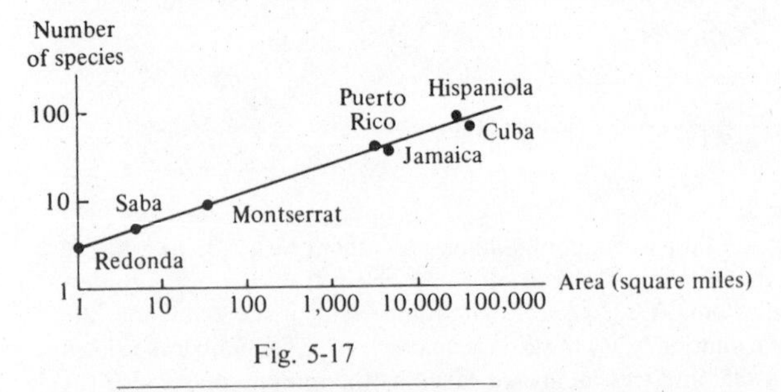

Fig. 5-17

The area-species curve of the West Indian herpetofauna (amphibians plus reptiles). (From MacArthur and Wilson, 1967.)

Habitat Islands on the Mainland

Some mainland habitats are obviously islands. Spruce bogs in deciduous forest, alpine mountaintops, and whole mountains separated by desert are obvious examples. In addition, a glance at actual land use and the vegetation of any region shows many more islands. Farmers' woodlots surrounded by fields, patches of second growth in old growth forest, recent fire burns, and so on, are all islands. To what extent, then, can we apply equilibrium theory to these mainland islands and get useful results?

The chief difference between mainland and oceanic islands is that the latter are separated by a vacuum insofar as land birds or insects are concerned, whereas mainland islands are separated by other habitats filled with birds and insects. Consequently, provided that the mainland islands are not too different from their surroundings, the birds from the surroundings are able to immigrate "onto" mainland islands more frequently and with more success than is true of immigration onto oceanic islands. There is no vacuum to be jumped on the mainland and dispersal over land is less hazardous to a bird species than dispersal over water. Thus, for example, for an island of spruce in the southeastern United States which is surrounded by deciduous forest and which is quite remote from other spruce islands, the immigration rate of deciduous forest birds (which can often survive in spruce forests) would be higher than that of spruce birds. This spillover of organisms from adjacent habitats is a primary factor present on mainland islands.

Brown (1971b) studied forested mountain islands in the desert lowlands of the Great Basin. These mountain islands have piñon-juniper woodlands at the base and conifer forests at higher elevations. The piñon-juniper zones show evidence of having been connected during recent geologic time. They have many relic species that are not being renewed by immigration, and so they have more than the equilibrium number of species. The higher conifer forests were never interconnected and have been colonized only by poorly adapted piñon-juniper mammals.

Vuilleumier (1970) has written a very interesting account of the bird species that occupy the mountain islands of the northern Andes. Figure 5-18 is a map of the area with the numbers of páramo bird

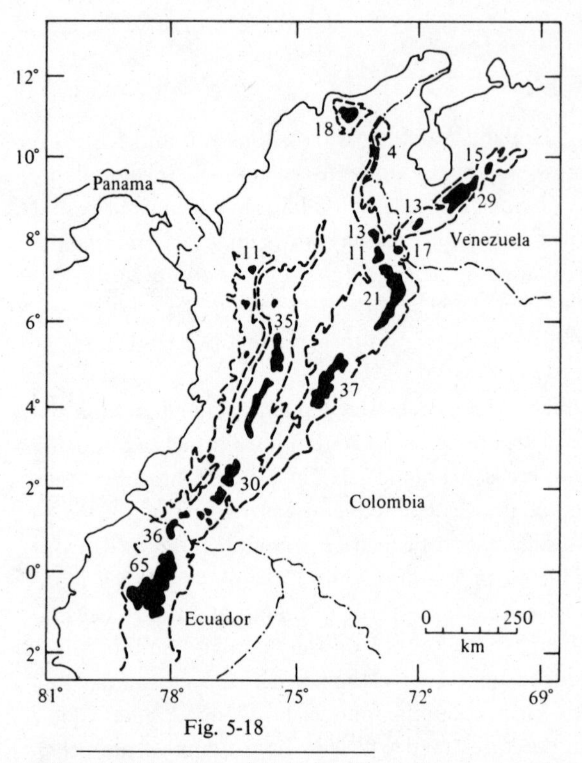

Fig. 5-18

The northern Andes, with the islands of páramo vegetation shown in black; the 1000-m contour is represented by the dashed line. The figures represent the total number of species on each island. (From Vuilleumier, 1970.)

species shown. It is immediately clear that both area and isolation are important. The mountaintops with páramo vegetation which lie near the main high Andes of Peru are quite rich in species, while the more remote and smaller "islands" are impoverished. Vuilleumier went through a sophisticated multiple regression and found that, compared to oceanic islands, the effect of distance is reduced on montane islands. Birds appear to colonize a distant mountain more readily than a distant oceanic island, and presumably the difference lies in the habitat to be crossed. A páramo bird that alights in lower-elevation forest can at least rest there and perhaps feed itself for a while (although it will not find enough food to feed a mate and young) whereas a transoceanic bird that gets tired is doomed.

If we viewed a mountain from above, its vegetation

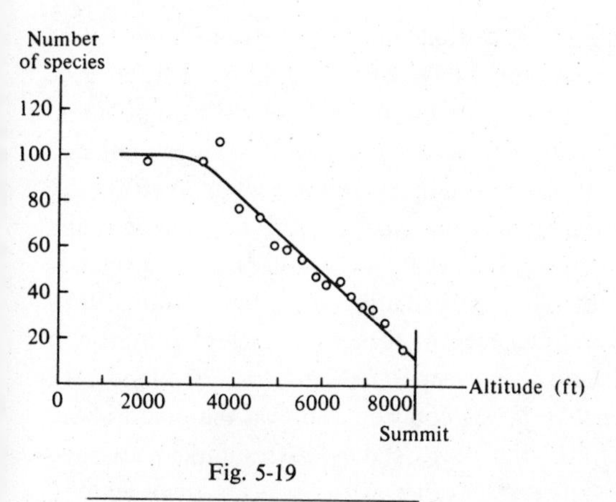

Fig. 5-19

Species diversity on Mt. Karimui, New Guinea. Each point gives the number of forest bird species recorded at the given elevation. Species diversity decreases with increasing altitude. (Provided by J. M. Diamond.)

zones would form concentric rings. The middle circle is the small mountain-top, the next ring the slightly larger area of the next lower forest type, and so on to the outermost ring, which has the largest area of all and is the lowest altitude belt. Since mountains are islands, we expect the smallest vegetation zones to have the fewest species. Not only are the highest vegetation zones smallest in area, but they are usually the simplest in structure, and often most remote and with the lowest immigration rate from similar habitats. For all of these reasons we expect the number of species to decrease with altitude, and, in fact, it does. Figure 5-19 shows this for a New Guinea mountain range. Diamond (in press) has further shown that lower mountain ranges, with smaller areas of each vegetation type, have fewer species in each vegetation type.

We make considerable use of mainland islands in discussing species distributions as affected by competition (pp. 137–140).

Land Bridge and Oceanic Islands

Islands can be roughly divided into two kinds: those that were recently connected to the mainland and whose colonists need not have crossed a water gap, and those that have never been connected to the mainland and whose species must have colonized across a water gap. We call those which were formerly connected by a land bridge "land bridge islands" and those that were never connected "oceanic islands." It is reasonably easy to infer which islands were connected when the ice of the most recent glaciation bound up much of the ocean water and lowered its level. During the most recent glaciation, the sea level was lowered a maximum of about 100 meters relative to the land (Flint, 1957). Hence, unless we have strong evidence of recent channel scouring or of silting, we infer that any island that is separated from the mainland by a channel less than 100 meters deep was connected to that mainland about 12,000 years ago. Such islands—the Pearl Islands, for example—are land bridge and recent ones at that. At the other extreme, deep-ocean islands, especially when volcanic, can usually be considered oceanic. The West Indies are surely of this type. There is a possible intermediate category of islands consisting of those which were connected in times previous to the recent glaciation but have been islands during and since that time. Any such earlier connection would have to be no more recent than the second most recent glaciation. We can conjecture that any vertebrate species that has been isolated that long from the mainland population would be classified as an endemic species, but there is no proof of this. In practice the classification turns out to be much clearer than we might fear, as we now show.

We compare large land bridge islands with oceanic islands first, and leave the interesting case of small land bridge islands until later. Using the 100-meter channel depth as a preliminary criterion of an oceanic island, we immediately find some striking results. Around the new world tropics, the West Indies, Cozumel off Yucatán, the Tres Marias Islands off western Mexico, and Curaçao and Bonaire off Venezuela are oceanic islands. Large land bridge islands in the tropics are Trinidad and Margarita off Venezuela, and Coiba and the Pearl Islands off Pacific Panama. There are many bird families found on these land bridge islands that are not found on the oceanic ones. For example, there are antbirds on Trinidad (9 species), Coiba (1 species), and the Pearl Islands (3 species), but there are no antbirds on the oceanic islands. This seems too regular to be due to chance, and we conclude that an antbird species on an island is a relic from the days of the land connection; for if it could colonize across water, we would expect it to be found on some oceanic island.

J. M. Diamond (pers. comm.) has carried out a more detailed analysis on the offshore islands of New Guinea. He lists 33 islands, or small groups of islands, 7 of which are large land bridge islands of over 200 square miles and 26 of which are truly oceanic. Of the New Guinea birds, 92 species are found on at least one of the land bridge islands but on no oceanic island, and 45 are found on three or more land bridge islands but on no oceanic island. Some rare species might have chanced to colonize one land bridge species across the water and yet never have happened to reach any oceanic island. Hence we cannot infer that all 92 species have remained from land bridge days, but there are two ways to reduce this figure to a likely number. First, at least the 45 species that are on 3 or more land bridge islands and missing from all oceanic islands must be relics; it would be very unlikely for a cross-ocean colonist to land on 3 of the 7 land bridge islands without landing on at least one of the more plentiful 26 oceanic islands. Another approach is to consider the species that are on one or more of the oceanic islands but on none of the land bridge islands. There are 19 of these. At most an equal number would be expected to have colonized some land bridge island and to have missed all the oceanic islands. Hence we subtract 19 from 92 to get 73 as a conservative estimate of the actual number of land bridge relic species. This estimate, however, does not tell us which species they are, so Diamond uses the 92 species as a generous estimate of the land bridge relics. On the large land bridge islands about $\frac{1}{3}$ of the species are apparently relics; that is, they are missing from all oceanic islands, while the remaining $\frac{2}{3}$ have colonized at least some oceanic islands across water and could have done the same to the land bridge islands. A similar analysis (MacArthur, Diamond, and Karr, in press) shows that about $\frac{1}{3}$ of the Pearl Archipelago species in Panama are probable relics of land bridge days.

Small land bridge islands are very interesting in this connection since they have an almost purely oceanic fauna. Thus while the large Pearl Island of Rey has about $\frac{1}{3}$ land bridge species, the small and remote Pearl Island of Contadora (Fig. 5-4) has one land bridge species among 20; $\frac{19}{20}$ are oceanic. In fact, even the single "land bridge" species, the flycatcher *Myiodynastes maculatus*, is doubtfully land bridge;

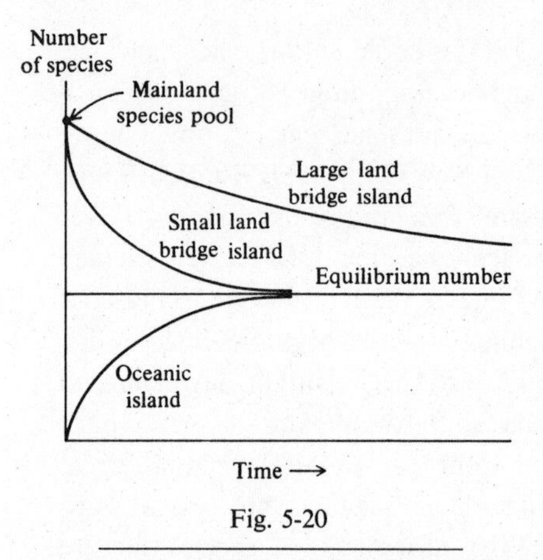

Fig. 5-20

The curves show how the number of species on oceanic islands and on small and large land bridge islands should change with time. The small land bridge island quickly loses its orginal fauna and converges to the oceanic island. The large land bridge island keeps many of its original species, and its number of species declines very slowly.

there are no oceanic islands off its range, so it was included as land bridge by default. Its semispecies, *M. luteiventris*, has colonized the oceanic islands of the Tres Marias (Grant, 1965), so perhaps all the current Contadora birds are oceanic. Neal Smith (pers. comm.) formerly found the land bridge barred antshrike (*Thamnophilus doliatus*) on Contadora, but it appears now to be extinct. The overwhelming impression one has on visiting a small land bridge island such as Contadora is just that of a small West Indian oceanic island. Contadora's common species are pigeons and doves, two sizes of hummingbirds, flycatchers of several sizes, bananaquits, honeycreepers, and vireos. This is a very West Indian family and species composition.

Examples of small land bridge islands in the New Guinea area with few or no land bridge species are the Schildpad Islands, Gemien, Pulu Adi, and the Madang Archipelago.

These observations are easily explained in terms of equilibrium theory and extinction patterns. Small land bridge islands, like small oceanic islands, have high extinction rates and rapid equilibration. They would be expected to retain no trace of their former connection. The extinction times of common birds on large oceanic islands are enormous, however, so these species will disappear from islands exceedingly slowly. If we could find a new oceanic island with no bird species but with a complete flora, and if we could find a newly severed land bridge island with complete mainland fauna and flora, we would expect the number of species on the oceanic island to rise and the number on the land bridge island to fall, both asymptotic to the same equilibrium level. But only a small land bridge island would complete the decay in a short time. Large land bridge islands retain their species for so long that we never expect to witness them at equilibrium (Fig. 5-20).

Island Evidence for Competition

So far we have described islands for their intrinsic interest. However, since they have fewer species than the same habitats have on the mainland, they can also be viewed as natural experiments. If we wish to know how the abundance of a species would change if we removed that species' competitors, we can look for an island which has the species present without its competitors. Purists often claim that these are rather poor experiments. They note that other conditions may also be different on the islands, and that these differences may be responsible for the contrasts in abundance. For instance, the islands may also be missing predators and our species may become commoner more by virtue of release from predation than by release from competition. So far, so good; we have an explicit and plausible alternative, and we must make the observations needed to discriminate between predator release and competitor release. But some critics go further and say that unspecified unexpected things may be different on the islands, and therefore that we should not be impressed by island-mainland comparisons. One could as validly say that unexpected things may be different between any experiment and its control and that experiments are not to be trusted. But island-mainland comparisons are really dramatic and call for some explanation. Let us examine some real cases.

A comparison between the previously mentioned island of Puercos in the Pearl Archipelago south of Panama and mainland plots of similar structure allows us to give a fairly detailed account of species released from competition. Puercos is 70 hectares in area and has only

20 of the mainland species, and these are mostly from the mainland second growth. The forest of Puercos is intermediate in structure between the two mainland plots used in this comparison: one plot moist forest and the other late shrub (Karr, 1971). On the basis of a year's observations, Karr judged that there were 56 and 58 species of birds resident on only 2 hectares of each of the two mainland plots and that larger areas would have had many more. Yet, abundances of the 16 resident forest interior island species were so high that there were 21.6 resident pairs per hectare compared to 18.2 and 14.2 for Karr's two mainland areas. In spite of the fact that there were fewer than half as many species on the island, there was a greater density of bird individuals, showing that the average abundance per species was much greater: 1.35 pairs per species per hectare compared to 0.33 and 0.28 pairs per species per hectare on the mainland. Something has allowed abundances of island species to increase! It is hard to compare particular species, because Puercos birds were from mainland second growth; even the "late shrub" had only 9 of the 20 Puercos species, and of those Karr only obtained significant population estimates of 6. For these 6 we give the Puercos abundance per hectare first, followed by the one for the mainland: white-tipped dove (*Leptotila verreauxi*), 24 vs. 12; red-crowned woodpecker (*Melanerpes rubricapilus*), 24 vs. 4; barred antshrike (*Thamnophilus doliatus*), 112 vs. 8; streaked flycatcher (*Myiodynastes maculatus*), 16 vs. 4; yellow-green vireo (*Vireo flavoviridis*), 104 vs. 35; red-legged honeycreeper (*Cyanerpes cyaneus*), 56 vs. 12. In these six cases the Puercos birds average 5.6 times commoner. There can be no doubt, then, that island species can increase greatly in density. Among these six, one—the woodpecker—is a hole nester. These are supposed to be much safer from predators because of the nest; in fact, their relative safety from predation is often given as the explanation for the greater number of eggs per clutch in hole nesters. But our woodpecker increased sixfold, which is fully as much as the predator-vulnerable species increased. Although the data are not so precise for the mainland, the other hole nester, the flycatcher *Myiarchus ferox*, also increased dramatically on the island.

This is a first line of evidence that the density increases are due to a release from competition rather than predation. But there is other evidence. First, the species that increased the most, the barred antshrike (which is a kind of antbird) is the species that lost the greatest number of mainland species of the same feeding habits. There were only two antbirds plus a wren gleaning low vegetation in Puercos, while nearly half of Karr's mainland species were antbirds. No wonder one antbird and the wren were the two commonest Puercos species!

Now, if the mainland species are restricted in abundance and habitat by competition, on the island we should expect not only an increase in abundance, but also in habitat and foraging height range. The latter finding was easy to document, and was the immediate impression of all the investigators. The proportion of foraging time was recorded in each of six height zones above the ground: 0–6 in., 6 in.–2 ft, 2–10 ft, 10–25 ft, 25–50 ft, >50 ft. The corresponding proportions of the species time were $p_1, p_2, p_3, p_4, p_5, p_6$, and a measure of the range of foraging height is $\frac{1}{\sum_i p_i^2}$. (It is instructive to contemplate this measure, which we shall see again in later chapters. p_1^2 is the probability that two separate birds of the species are simultaneously in height zone 0–6 in., p_2^2 is the probability that both are in the second layer, and so on. Thus $\sum_{i=1}^{6} p_i^2$ is the probability that both are feeding in the same layer, no matter which layer it is. This is large if the species has restricted feeding height, so that both individuals are likely to be feeding together; thus $\sum_i p_i^2$ increases with the narrowness of the belt of foraging heights. Its reciprocal, $\frac{1}{\sum_{i=1}^{6} p_i^2}$, increases with the width of the foraging height. If, in fact, the bird spends its time equally in three layers, say, 1, 2, and 3, then $p_1 = p_2 = p_3 = \frac{1}{3}$ and $\frac{1}{\sum p_i^2} = \frac{1}{\frac{1}{9} + \frac{1}{9} + \frac{1}{9}} = 3$. In other words, if a species feeds equally over n layers, the measure $\frac{1}{\sum_i p_i^2}$ gives us the number n of layers. It is often interpreted as the number of "equally used layers." See Horn (1966).) G. Orians (pers. comm.) in Costa Rica had already measured foraging heights of two of the Puercos species, and we give the Puercos number of foraging layers first followed by the Costa Rica mainland value: bananaquit (*Coereba flaveola*), 2 vs. 1.33; yellow-green vireo (*Vireo flavoviridis*), 2.09 vs. 1.78. There are also mainland measurements for two close relatives of Puercos species: the wren *Thryothorus leucotis*, 3.01 (Puercos) vs. the wren *T. pleurosticticus*, 2.55 (Costa Rica); the antshrike *Thamnophilus doliatus*, 3.95 (Puercos) vs. the antshrike *T. punctatus*, 1.24 (Costa Rica). In all cases the island birds have expanded their range of foraging height just as they have expanded abundance. Expansion is easy to predict from competitive release but it is hard to predict from other premises.

To document habitat expansion, it is sufficient to note that islands of very different vegetation structure often contain the same bird species. On the mainland those species would be very much more restricted.

Not only do island species enlarge their foraging ranges and habitats, but they also use a clever device to increase their range of food size. We know food size is closely determined by the forager's morphology and the menu of available food items and is, therefore, not easily altered by absence of competitors. However, often the different sexes are of different sizes on islands, so that in combination they eat a wider range of foods. And even where they are not of different size, the sexes may eat different food sizes. Figure 5-21 shows the sizes of food taken by males and females of the lizard *Anolis conspersus* on Grand Cayman Island, as determined by Schoener (1967). Clearly they have widened their utilized resource spectrum.

Finally, there are what Diamond calls checkerboard distribution patterns on islands, in which two or more closely related

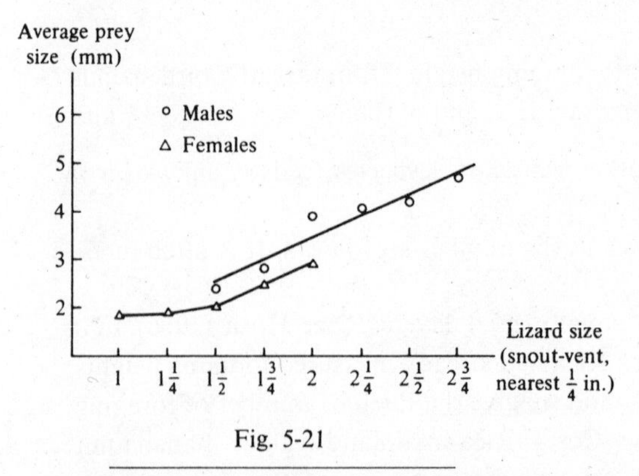

Fig. 5-21

Relationship of predator size to prey size for male and female lizards. The lizard is *Anolis conspersus* from Grand Cayman Island. (From Schoener, 1967.)

species occupy similar islands to the exclusion of each other, but in a geographically irregular pattern. Thus some dove of the genus *Leptotila* is on virtually every island near Central America. *Leptotila verreauxi* is not only on Puercos and the rest of the Pearl Islands but also on Taboga, Urava, the Tres Marias, Aruba, Bonaire, and Curaçao. On the island of Coiba off Pacific Panama, however, *L. battyi* is present and *L. verreauxi* has never colonized. Again, no Panamanian island seems to have more than one abundant wren, but it is *Thryothorus nigricapillus* on Escudo de Veraguas; *T. leucotis* on Rey, Puercos, and Viveros; and *Troglodytes musculus* on nearby San Jose, Pedro Gonzales, and Trapiche, and also on Coiba. It is hard to avoid the conclusion that these checkerboard patterns are due to competition. Lack (1971) gives many other examples in West Indian birds (Fig. 5-22).

Wilson and Taylor (1967) have given another example of checkerboard patterns, this time in ants of Polynesia, where large aggressive

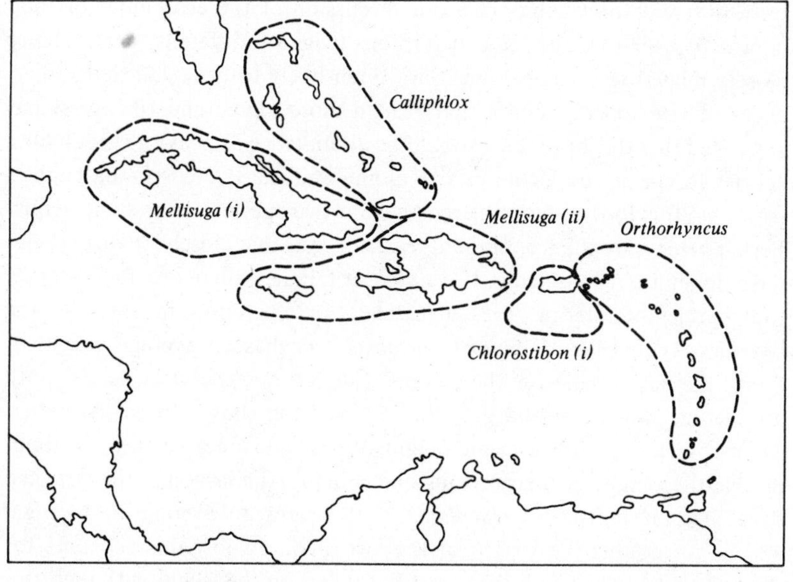

Fig. 5-22

Geographical replacement of small hummingbirds in West Indies. There is only one small species on each island, derived from one of four separate genera. (From Lack, 1971.)

ants of the genus *Pheidole* are involved. *Pheidole fervens* is dominant in the Society Islands, on Tonga, and on Pitcairn Island; *P. megacephala* is dominant on Upolu in Samoa, on the Marquesas, and in Hawaii; *P. oceania* replaces *P. megacephala* on Savai'i in Samoa. In the Dry Tortugas of Florida and on the small islands off Puerto Rico *P. megacephala* is again involved in a checkerboard, this time with the fire ant *Solenopsis geminata*.

There is one other aspect of competition, or potential competition, among Puercos bird species that deserves mention. In general, an island immigrant is less likely to colonize successfully if it has a close competitor already present. We pointed this out in the section on colonization, and it is the explanation for the checkerboard patterns we have just discussed. This competition has another effect. We have seen how island species expand their vertical foraging range. This means a new colonist is less likely to be able to survive if it only differs on the mainland in foraging position. New colonists that differ in size, however, still should have different food habits (see our discussion of the economics of food choice, pp. 60–61) and can perch on twigs of different sizes. Hence closely related species on islands are often likely to have sufficient differences in size to coexist for that reason alone. Diamond (in press) has suggested that the larger bird should be about twice as heavy as the smaller species to coexist by virtue of size alone, and the Puercos fauna enables us to test this too. In fact, Puercos has no two species of the same genus! Perhaps this is by chance, but it at least suggests that closely related species have difficulty in colonizing. If we look at families, there are, for instance, four species of interior forest flycatchers on Puercos. The smallest, the beardless tyrannulet, *Comptostoma obsoletum*, has an average weight of 8 gm; the next smallest is the scrub flycatcher, *Sublegatus arenarum*, with an average weight of 14.6 gm; then follows the short-crested flycatcher, *Myiarchus ferox*, with a mean weight of 33.3 gm. Each of these is about double the weight of the previous one. Finally, the largest is the streaked flycatcher, *Myiodynastes maculatus*, that weighs an average of 44.5 gm and is not double the weight of the last. (Conceivably this accounts for the fact that it was much the rarest flycatcher on the island, although this could have other explanations.) Other families do not seem to sort by size. For instance, there are two flower feeding "honeycreepers" (probably only distantly related phylogenetically), the bananaquit (*Coereba flaveola*) and the red-legged honeycreeper (*Cyanerpes cyaneus*) on Puercos. They feed together among the flowers in the canopy and their mean weights are 10.7 and 12.8 gm, respectively. There is a plausible explanation for the difference. Large flycatchers do eat larger foods than small ones (Hespenheide, 1971) and so could coexist by size difference, whereas there is no simple way that a large honey eater could eat from different food than a small one. Conceivably, it could be harmed by being able to perch only on stronger twigs, but it seems likely that coexistence is not based on size. Rather the bills are of quite different shape, and it is very likely that these species eat nectar from different flowers or eat different insects while feeding on nectar. Snow (1962) has documented that they forage for insects in very different ways.

We have included the rather long case history of Puercos to illustrate just what can be learned from island-mainland comparisons, but it is by no means unique. Crowell (1962) and Grant (1965) reached similar conclusions comparing Bermuda and the Tres Marias islands to their respective mainlands. Diamond, on the other hand, in brilliant studies of the islands off New Guinea, duplicated only some of the Puercos results. On the island of Karkar, he documented niche shifts. Of the island's 52 species, 23 underwent no niche shifts, and 22 expanded spatially by occupying a greater range of habitats (12 species), altitudes on mountainsides (13 species), or foraging heights (5 species); in addition, 12 species became denser on the island (in their preferred mainland habitat), a single species expanded its diet, and one expanded its foraging technique. These are similar to the Puercos results though more detailed in some respects. But Diamond found the mean abundance per species no greater on the island than on the mainland in spite of all these expansions. He suggests that the islands off New Guinea were colonized by species that were very inappropriate to their islands, while Puercos was colonized by birds that were much less inappropriate and hence could increase in density. The appropriateness of the colonists is related to the proximity and abundance of the island habitat on the mainland and to history. Land bridge islands presumably had a better chance to try mainland species for suitability. In any case, although all island studies show some niche expansions, only some show expansions of abundance.

The Krebs Effect

One of the most remarkable results of recent experiments in ecology was reported by Krebs and his colleagues (Krebs, Keller, and Tamarin, 1969). They enclosed 2-acre fields with wire mesh fences and topped them with sheet metal so that no mouse could enter or escape. The main occupants of the enclosures and of the unenclosed control fields were mice of the genus *Microtus*, *M. pennsylvanicus* and *M. ochrogaster*. What is astonishing is that the mouse populations in the enclosures quickly grew to a very much higher level than the populations in the similar but unenclosed fields. In fact, the combined populations of the two *Microtus* species in the enclosed fields reached 3 times the density of the unenclosed plots. Krebs believes that the population increase in the enclosures cannot be attributed to the lack of predation since the predators were not effectively fenced out; for instance, he noted fox tracks in the snow crossing his fences. Also, the shrew *Blarina*, which certainly eats some *Microtus*, was fenced in with the mice. But this point needs further confirmation. Instead, Krebs conjectures that the fences in some way interfered with the normal population regulation machinery of the mice, perhaps by preventing the emigration of young mice. The exact cause of the greater population density in the enclosures is still a mystery, but this result is of considerable importance in population dynamics. For our purposes, it is the effect itself that is of interest: enclosed populations, of rodents at least, often reach higher levels than unenclosed populations.

Clearly this effect could be important on islands, which are enclosed, and might account for some of the island increase in population density. Two general points in this connection must be made. First, an enclosure the size of North America (in fact, North America is enclosed by oceans) does not have the same effect as the 2-acre enclosure. Thus the effect must virtually disappear as the enclosure, or island, gets bigger; but the size of the island needed virtually to eliminate the effect is unknown. If adjacent small and large islands have equal population densities, we can presumably rule out the Krebs effect, while if the small one reaches much higher density, the Krebs explanation remains plausible. As an example, preliminary evidence for birds on the Pearl Islands suggests that densities are about equal on Puercos (70 hectares) and on Rey (well over 20,000 hectares), so we conclude very tentatively that the Krebs effect has not played much role in bird population changes on the islands. However, both Puercos and Rey have higher total bird densities than the mainland. Here the Krebs effect may explain the disparity.

The second point is that Krebs' conjectured explanation is, in fact, most plausible for grass-eating herbivores. Such herbivores might easily devastate their food supply, and a population-regulating mechanism that prevents this might well be favored. Birds could also profit from mechanisms to regulate population since many have a fixed, nonrenewing food supply of fruits and seeds in the winter. They could cause their own extermination in February by having too many birds harvesting in December.

Whatever the ultimate explanation of the Krebs effect may prove to be, it can undoubtedly be observed at times on islands.

The Island View of Competition

Local extinction is an added ingredient of habitats that occur in isolated patches, or islands. It means that some patches apparently suitable for occupation will be unoccupied, and this, in turn, has profound implications for competition. If, for instance, two species of competitors try to occupy an area, each will be found in scattered colonies. Individuals of each species will therefore more often be with their own species than with the competitor, and each species will thus inhibit itself more than it will the other. As a result, they may coexist even though in a continuous habitat one might always outcompete the other. For theoretical calculations along these lines see Skellam (1951) and Levins and Culver (1971). Levins and Culver point out further that a species may be kept out of the area even by a very rare competitor. For this to occur, the species' extinction rate from patches of environment would have to almost equal its rate of immigration, so that its existence would be already precarious. Then even a rare competitor might tip the balance, and extinction plus competitive elimination would exceed immigration.

We can extend this reasoning to patches of two habitat types, of the sort that species 1 does best in patches of type 1 and species 2 in patches of type 2. Suppose patches of type 2 are just far enough apart so that migration from one to another is less than extinction. Then species 2 will only persist if it can also occupy patches of type 1, thereby increasing its rate of immigration. Therefore, we may find a situation such that although each species is superior in its own kind of habitat, species 1 outcompetes species 2 by preventing 2 from occupying the necessary patches of type 1.

These examples offer a sample of the applications of the island view of competition. Within each patch of habitat, the outcome of competition is as described in Chapter 2, but in an entire area one species may eliminate another, or coexist with it, because of immigration and extinction rates, and independently, in part, of competitive ability.

Appendix *Extinction Probability*

There is an elegant way of showing the probability that the descendants of a founder will go extinct. We let λ be the instantaneous per capita birth rate so that an individual has probability $\lambda\, dt$ of having an offspring in very short time dt. Thus a population of m individuals has probability $m\lambda\, dt$ of having one offspring in time dt. With dt so short, there is negligible probability of more than one birth. Similarly, μ is the instantaneous per capita mortality so that the population of size m has probability $\mu m\, dt$ of losing one in the very short time dt. The remaining probability, $1 - \lambda m\, dt - \mu m\, dt$, is the likelihood that the population of size m is still of size m after time interval dt. The fact that we let λ and μ be constants means that these rates do not change with population density; we are describing a population that *expects* to grow exponentially according to $\frac{1}{X}\frac{dX}{dt} = \lambda - \mu$ or $X = X_0 e^{(\lambda-\mu)t}$. But it can go extinct if by chance a great many mortalities occur before the births. Now suppose the population is currently of size m. Consider the function $\left(\frac{\mu}{\lambda}\right)^m$, whose virtues will appear presently. The expected value of this function after time dt is easy to calculate: The population will have size $m + 1$, so that the function is $\left(\frac{\mu}{\lambda}\right)^{m+1}$, with probability $\lambda m\, dt$; the population will have size $m - 1$ and the function takes the value $\left(\frac{\mu}{\lambda}\right)^{m-1}$, with the probability $\mu m\, dt$; and the population will remain m and the function remains $\left(\frac{\mu}{\lambda}\right)^m$ with probability $(1 - \lambda m\, dt - \mu m\, dt)$. The expected value of the function after time dt is thus $\lambda m\, dt\left(\frac{\mu}{\lambda}\right)^{m+1} + \mu m\, dt\left(\frac{\mu}{\lambda}\right)^{m-1}$ $+ (1 - \lambda m\, dt - \mu m\, dt)\left(\frac{\mu}{\lambda}\right)^m = \mu m\, dt\left(\frac{\mu}{\lambda}\right)^m + \lambda m\, dt\left(\frac{\mu}{\lambda}\right)^m + \left(\frac{\mu}{\lambda}\right)^m$

$- \lambda m\, dt\left(\frac{\mu}{\lambda}\right)^m - \mu m\, dt\left(\frac{\mu}{\lambda}\right)^m = \left(\frac{\mu}{\lambda}\right)^m$, the same as the value before the interval *dt*. Repeating this argument we see that the expected value of the function does not change with time. (This is what mathematicians call a martingale.) Suppose we begin with a propagule of one female or pair, so that the function begins with the value $\left(\frac{\mu}{\lambda}\right)^1 = \frac{\mu}{\lambda}$ and we know that the expected value of the function at any future time is also $\frac{\mu}{\lambda}$. Suppose we wait so long that the population has either gone extinct, with probability Ext, or grown to an enormous size (virtually infinity), with the remaining probability 1 − Ext. At that time the expected value of the function is therefore Ext $\left(\frac{\mu}{\lambda}\right)^0 + (1 - \text{Ext})\left(\frac{\mu}{\lambda}\right)^\infty = \text{Ext} + 0 =$ Ext, since $\mu < \lambda$ so $\left(\frac{\mu}{\lambda}\right)^\infty = 0$. We know this expected value equals the initial value, so Ext $= \frac{\mu}{\lambda}$ if we begin with one individual female (or pair). If we begin with a propagule of two (pairs), the function begins with the value $\left(\frac{\mu}{\lambda}\right)^2$, which equals Ext, the extinction probability, and so on. The probability of one individual (pair) leaving a successful colony is then $1 - \frac{\mu}{\lambda} = \frac{\lambda - \mu}{\lambda} = \frac{r}{\lambda}$, and the probability that a colony started by j pairs will be successful is $1 - \left(\frac{\mu}{\lambda}\right)^j = \frac{\lambda^j - \mu^j}{\lambda^j}$. If, for instance, $\lambda = 2$ and $\mu = 1$, then a single pair will go extinct with probability $\frac{1}{2}$ and will set up a large colony with probability $\frac{1}{2}$. Three such pairs will leave an extinct line of descent with probability $\left(\frac{\mu}{\lambda}\right)^3 = \frac{1}{8}$. With probability $\frac{7}{8}$ they will successfully colonize.

We can also ask the probability, 1 − Ext, that the population reaches size K before it goes extinct. If founded by a single pair, we must then have $\left(\frac{\mu}{\lambda}\right)^1 = \text{Ext}\left(\frac{\mu}{\lambda}\right)^0 + (1 - \text{Ext})\left(\frac{\mu}{\lambda}\right)^K = \text{Ext}\left(\frac{\mu}{\lambda}\right)^0$

$- \text{Ext}\left(\frac{\mu}{\lambda}\right)^K + \left(\frac{\mu}{\lambda}\right)^K$, whence Ext $= \dfrac{\frac{\mu}{\lambda} - \left(\frac{\mu}{\lambda}\right)^K}{1 - \left(\frac{\mu}{\lambda}\right)^K}$. But if K is reasonably large, $\left(\frac{\mu}{\lambda}\right)^K$ is negligible and we get our former result, Ext $= \frac{\mu}{\lambda}$. So a species is safest if $\frac{\lambda - \mu}{\lambda}$ is as large as possible; such a species is a good colonist.

A population that cannot exceed level K does not have the possibility of reaching such size that it is safe from extinction. All such populations must eventually go extinct, but some may expect to last a very long time before the inevitable occurs. For a population that expects to grow with per capita birth rate λ and per capita mortality μ until the population reaches level K, at which point no further growth is possible because a population of size $K + 1$ is suddenly reduced to size K, it is easy to calculate the expected time for the line of descent of a single pair of colonists to go extinct. Let T_j = the expected time for a population now of size j to go extinct. We know this is the time preceding any population change plus T_{j+1} times the probability the first population change is to population $j + 1$, plus T_{j-1} times the probability the first change is a decrease to $j - 1$. The expected time preceding any change is the reciprocal of the expected changes per unit time $= \frac{1}{(\lambda + \mu)j}$. The probability that the first change is an increase is $\frac{\lambda}{\lambda + \mu}$ while the probability that the first is a decrease is $\frac{\mu}{\lambda + \mu}$. Hence we can write

$$T_j = \frac{1}{(\lambda + \mu)j} + \frac{\lambda}{\lambda + \mu} T_{j+1} + \frac{\mu}{\lambda + \mu} T_{j-1} \tag{1}$$

This equation holds for all j up to $j = K$. For $j = 1$ we have a $T_{j-1} = T_0$ which is of course zero. For $j = K$ we have a T_{K+1} which we set equal to T_K assuming that the second the population reaches $K + 1$ it is reduced to K. Hence we have K equations in the K unknowns T_j ($j = 1, \ldots, K$) whose solution for T_1 can be calculated by computer

(Fig. 5-8). If we write the left-hand T_j in Eq. (1) as $\frac{\lambda}{\lambda+\mu} T_j + \frac{\mu}{\lambda+\mu} T_j$, the equations take the form $\frac{\mu}{\lambda+\mu}(T_j - T_{j-1}) = \frac{1}{(\lambda+\mu)j} + \frac{\lambda}{\lambda+\mu} \times (T_{j+1} - T_j)$ and the $\frac{1}{\lambda+\mu}$ cancels so $T_j - T_{j-1} = \frac{1}{\mu j} + \frac{\lambda}{\mu}(T_{j+1} - T_j)$. In particular

$$T_1 = T_1 - T_0 = \frac{1}{\mu} + \frac{\lambda}{\mu}(T_2 - T_1) \tag{2}$$

and $T_2 - T_1$ is given by $T_2 - T_1 = \frac{1}{2\mu} + \frac{\lambda}{\mu}(T_3 - T_2)$. Substituting this into (2), $T_1 = \frac{1}{\mu} + \frac{\lambda}{\mu}\left[\frac{1}{2\mu} + \frac{\lambda}{\mu}(T_3 - T_2)\right] = \frac{1}{\mu} + \frac{\lambda}{\mu}\frac{1}{2\mu} + \frac{\lambda^2}{\mu^2}(T_3 - T_2)$. Then we substitute the formula for $T_3 - T_2$, and so on. Carrying this out until we have $T_{K+1} - T_K$, which is zero, in the right side, gives us

$$T_1 = \frac{1}{\mu} + \frac{\lambda}{\mu}\frac{1}{2\mu} + \left(\frac{\lambda}{\mu}\right)^2 \frac{1}{3\mu} + \cdots + \left(\frac{\lambda}{\mu}\right)^{K-1} \frac{1}{K\mu} \tag{3}$$

which can also be written in the form

$$T_1 = \frac{1}{\lambda}\frac{\lambda}{\mu} + \frac{1}{2\lambda}\left(\frac{\lambda}{\mu}\right)^2 + \frac{1}{3\lambda}\left(\frac{\lambda}{\mu}\right)^3 + \cdots + \frac{1}{K\lambda}\left(\frac{\lambda}{\mu}\right)^K \tag{3'}$$

The last term $\frac{1}{K\lambda}\left(\frac{\lambda}{\mu}\right)^K$ is larger than the others and can be taken as an (under)approximation of T_1. Now we know that a fraction equal to $\dfrac{\frac{\mu}{\lambda} - \left(\frac{\mu}{\lambda}\right)^K}{1 - \left(\frac{\mu}{\lambda}\right)^K}$, which is in turn approximately equal to $\frac{\mu}{\lambda}$, of the immigrant pairs will fail even to establish a colony of size K, and its expected time to go extinct in any case is slightly greater than $\frac{1}{K\lambda}\left(\frac{\lambda}{\mu}\right)^K$. As the figure shows, this time can be small, or, with large K, exceedingly large.

Suppose now that a species invades two islands that are

identical except that one has a competitor of $\alpha = 0.5$. Without the competitor our species reaches a population asymptote of K and with the competitor it reaches $\frac{K}{1+\alpha}$, if both have the same K; for then both species have the same equilibrium populations X and obey the equations $0 = \frac{1}{X}\frac{dX}{dt} = \frac{r}{K}[K - X - \alpha X]$. Hence on the island with the competitor, our species will asymptote at the population level $\frac{K}{1+\alpha} = \frac{2}{3}K$. $\frac{\lambda}{\mu}$ will also be reduced on the island with the competitor. In fact, initially $\lambda - \mu = \frac{1}{X}\frac{dX}{dt} = \frac{r}{K}[K - \alpha K] = r(1-\alpha)$ where the competitor is present, so $\lambda - \mu$ is reduced from r to $r(1-\alpha)$ where the competitor is present, but it is not quite clear how much $\frac{\lambda}{\mu}$ will be reduced. It can be written as $\frac{r+\mu}{\mu} = 1 + \frac{r}{\mu}$, and r is reduced to $r(1-\alpha)$ but we have not specified how μ will change. If only λ changes and μ stays constant where there is a competitor, then $T_1 \sim \frac{1}{K\lambda}\left(\frac{\lambda}{\mu}\right)^K$ and $\frac{1}{K\mu}$ is constant. To take a specific example, we further let $K = 20$ and $\mu = 1$ so that $\lambda = 2$ without competitors ($r = 1$) and $\lambda = 1.5$ with competitors leaving a new rate of increase $r(1-\alpha) = 0.5$. Then on the island without competitors the expected survival time of a colony founded by one pair is approximated by $\frac{1}{K\lambda}\left(\frac{\lambda}{\mu}\right)^K = \frac{1}{40}2^{20}$, which is about $\frac{1{,}000{,}000}{40} = 25{,}000$ years. With the competitors the same species would expect to last only $\frac{1}{13.3 \times 1.5}(1.5)^{13.3}$, which is about 11 years. In this case the competition has reduced survival time from 25,000 years to 11 years! The reduction would be even more drastic if μ, rather than λ, were altered by competition. Then, to have $r = 0.5$ we would have $\lambda = 2$ and $\mu = 1.5$, whence $\frac{\lambda}{\mu} = 1.33$. The new survival time would be $\frac{1}{K\lambda}\left(\frac{\lambda}{\mu}\right)^{K} = \frac{1}{26.6}(1.33)^{13.3} = 1.2$ years. Different λ, μ, and K values would of course alter the calculation, but the effect of the competitor on survival is very dramatic. Just the reduction of the exponent K to $\frac{2}{3}K$ makes the survival time the $\frac{2}{3}$ power of its former time. That factor alone would reduce 20,000 years to 700 years or reduce 500 years to 63 years. And the reduction in $\frac{\lambda}{\mu}$ magnifies the reduction in survival time with competition. This is why colonists rarely succeed on an island with a competitor of α as large as 0.5.

Species Distributions

6

The ranges of single species would seem to be the basic unit of biogeography. Curiously enough, the history of science often proceeds in a reverse order from expectations and this is very true of biogeography. Patterns on islands (Chapter 5), of species diversity (Chapter 7), and of tropical communities (Chapter 8) are already clear and even moderately well understood, while patterns of single species' ranges still seem to be catalogs of special cases. In this chapter an attempt is made at classifying kinds of range boundaries and the kinds of flow of populations that take place within those boundaries, but no very pleasing pattern emerges.

When the ranges of many species are superimposed we are faced with "community" patterns and can ask new questions, chiefly concerned with the degree of synchrony in the appearance and disappearance of species. Such questions occupy us in the second part of the chapter.

Single Species

We approach the matter of single-species distributions by considering first some of the limitations imposed by climate. We proceed then to the various ways species might meet the problem of unpredictability of climate.

Adaptations to Climate

A host of plant species can occur only where it never freezes, and these are confined to the warm tropics. The saguaro cactus (*Cereus giganteus*), that large succulent cactus famous in Arizona advertisements, cannot stand long freezes. More specifically, 36 hours of temperature below freezing seem to kill many of them. Where it thaws every day, the saguaro won't freeze, and the northern and eastern edges of the saguaro distribution correspond to the line bounding places where on occasional days it does not thaw (Fig. 6-1). The southern edge of the saguaro's range in Mexico has no such simple climatic explanation, but as southern edges of ranges pose special problems, we only note this fact now and discuss southern edges in a special section.

Plants have moisture limitations, too. The same pleated saguaro cactus appears unable to survive long with its roots in standing water, but it does need occasional rains, which it stores by expanding accordion fashion. Saguaro is much more efficient at storing warm

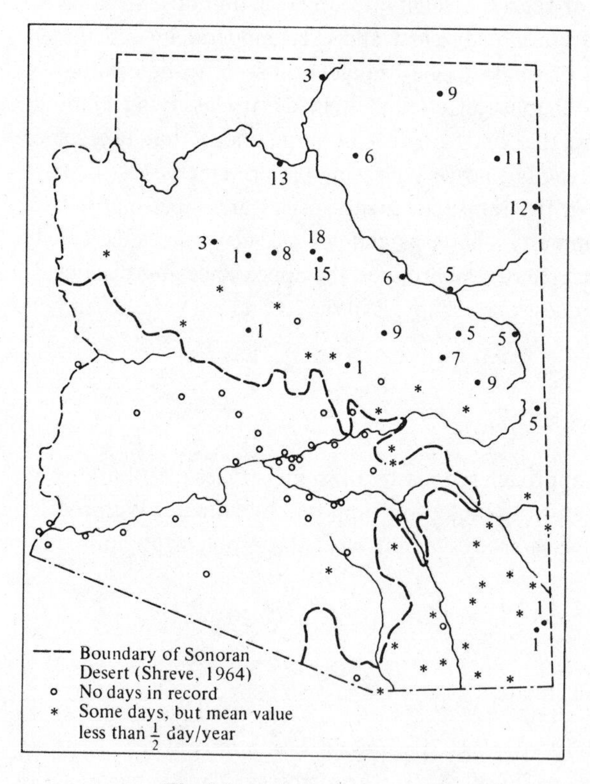

Fig. 6-1

Northern and eastern edges of the Sonoran Desert, where the saguaro cactus lives, correspond closely to the line beyond which it occasionally fails to thaw. The numbers are mean number of days per year with no rise above freezing temperature. (From Hastings and Turner, 1965.)

summer rains than cold rains, which may explain why the winter rain deserts of southern California have no saguaros (Hastings and Turner, 1965). Anyone can supply examples of water-loving plants whose very way of life would preclude their living on dry land; for example, water lilies and duckweed.

Examples of simultaneous control of plant distribution by temperature and rainfall (or other factors acting simultaneously) are common but are less well worked out. In Great Britain the moor rush, *Juncus squarrosus*, is found at all elevations up to nearly 3000 ft, although at higher elevations its size and number of flowers diminish and above about 2200 ft few of the flowers mature into seed capsules. A few plants reach 2700 ft (Pearsall, 1950). This distribution is due to the length of the warm season (Fig. 6-2). In 1942, at 700 ft elevation, flowering was completed during June; at 2000 ft, flowering did not begin until after the end of July; and between 2500 and 3000 ft, flowering was not completed by the end of August. (These results were for the Lake District in 1942 but similar patterns have occurred at different times and places.) The rush does not exist higher on the mountains than where it can normally produce seed. Perhaps at very rare intervals, during an exceptionally warm summer, seeds are produced at higher elevations and the long-lived plants are survivors of this occurrence.

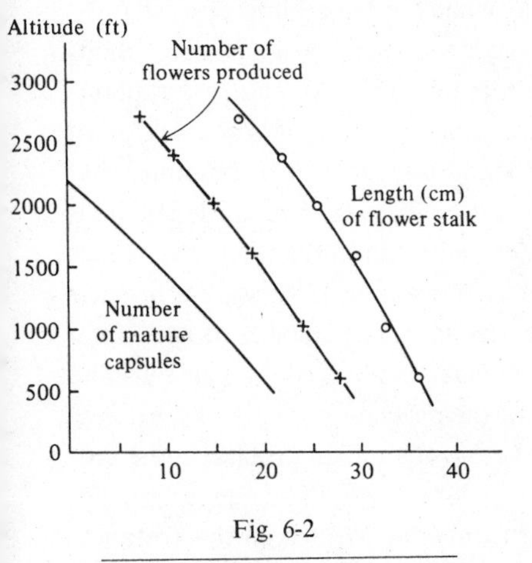

Fig. 6-2

Effect of altitude on moor rush, *Juncus squarrosus*. (From Pearsall, 1950.)

Pearsall also describes the tiny rush moth (*Coleophora caespititiella*), whose larvae feed on the growing seeds inside the ripening fruit of the moor rush. The eggs are laid on the flowers of the rush only in June and July; hence no moths could be found above about 1800 ft because no rushes were in bloom during those months above that elevation. Pearsall believes there may also be a minimum temperature for normal larval development, so that any attempt to postpone egg laying to coincide with the later rush flowering of higher elevations would be unsuccessful. In any case, the example shows the kind of interaction between presence of appropriate plants and of climate that so often controls animal distributions.

Another well-worked-out case is the climatic control of the alpine sorrel, *Oxyria digyra*, in studies by Mooney and Billings (1961). The book *Silvics of the Forest Trees of the United States*, published by the U.S. Department of Agriculture, lists temperature and rainfall limits for the present geography of important tree species.

Adaptation to Unpredictable Climate

It is simple to say that such-and-such a species is adapted to a cold winter or to a dry summer. In very few cases do we know the machinery of such an adaptation, and when we have hints of it, it turns out to be complicated. Horticulturists, concerned with growing plants outside of their normal range, have seen many of the difficulties that plants face. They have coined the term "hardening" for a plant's acquisition of tolerance to winter; the plant cannot stand any frost in the summer, but by, say, mid-October it becomes resistant to frost—it has hardened. This makes early and late frosts of greater importance than normal winter ones. Peaches grow well in the Niagara Peninsula of Ontario, not because it is so warm there but because spring is so late. The chances of accidental frosts are very low by the time the climate, moderated by nearby Lake Ontario, warms enough in the spring for the peaches to lose their hardiness and send out buds. Farther south, and far from a lake, peaches come into bud earlier and hence are vulnerable to late frosts. Thus in a warmer climate peaches may more often suffer from cold. We see, in this instance, how important the unpredictable storm can be.

Similarly, in the deserts the irregularity of rainfall is often as important as its dearth. A desert annual plant only lives one season and must in that season produce seeds to wait dormant until the next suitable rainy season. If all the seeds germinated after the first rain in a year with no subsequent rains, they might all wither before setting seed, and the plant would go extinct locally. To prevent this, plants have adopted various strategies. Some have a fraction of their seeds that do not germinate the first year. If that year proves too dry, some seeds are left, then, to try the next, and later, years. Cohen (1967) has worked out mathematically what fraction of the annual seeds should germinate, and as one would guess, the more probable a bad year with insufficient rain, the more seeds will delay their germination. Other plants have some sort of rain gauge and don't germinate until a great deal of rain has fallen. By this time the ground may be wet enough. For instance, the palo verdes (*Cercidium* sp.) of Arizona seem to have a germination inhibitor that must get scrubbed off when the seed tumbles in the sand down a desert wash in a summer storm. This ensures a fair water supply, although the palo verde, as a tree, is much safer than an annual plant; if 1 year's seed supply fails, the tree can produce more the following year.

A similar problem faces a small rodent (say a pocket mouse of the genus *Perognathus*) deciding whether or not to hibernate. Of course, the mouse does not consciously decide to hibernate. Evolution has to decide, which it does by its usual trial-and-error procedure. Those mice that leave a great many descendants have their type perpetuated; our mouse's genetic constitution presumably causes the mouse of one location to hibernate, while a different genotype in a different climate may cause the mouse to ignore the same signals of approaching winter. Presumably, if all winters were uniformly severe, all mice would hibernate; or perhaps none would. But if some are severe and others are mild, and there is no predicting in advance which is coming, what should the mouse do? The no-hibernating genotype would have the advantage in a mild winter by maintaining its full body mass and perhaps even reproducing, but on the other hand, all of these would die in a severe winter. As with the desert annuals, the best strategy is for some to do each. This example is purely hypothetical, however. *Perognathus* apparently hibernate in the northern plains and Canada and not in southern Arizona and Mexico, but the proportion hibernating is unknown for any locality.

The Failure of Climate Alone to Account for Distributions

It would formerly have seemed to many people that this chapter would be the longest in the book—on the assumption that most plants occur just where their climate needs are met. Of course, if this were strictly true, flower gardens and arboretums would be impossible; every plant that could grow in a particular place would already be present. Faced with the fact that gardens and arboretums really are feasible and can grow vast numbers of plants not naturally occurring in the region, we can give two possible reasons. The first that occurs is that history hasn't afforded those plants the chance and that if they had been introduced they would have taken their place among the local wild plants. At first this seems to account for virtually all missing plants, but on further thought it proves inadequate. If only lack of introduction were the cause of the absence of viable plants, the plants in our gardens would never need weeding! Once introduced, they would persist of their own accord. Another way to view the question is this: Very many European plants have been introduced accidentally to North America, but relatively few have invaded the native woodland vegetation; the vast majority occupy cultivated areas and roadsides—new habitats that man introduced to North America and that had no native eastern plants of their own. There are many European tree species growing and reproducing in the cities, where they are "weeded" by having the lawns about them cleaned of competitors; how many of these have we seen growing in the natural forests? If the answer is only few; it is for nonclimatic reasons.

The Role of Competition

Southern edge of ranges. There is another frequent pattern that seems not to be explainable in terms of climate alone. From about central United States on south toward the equator, summer climates actually become more equable. Not only are winter temperatures much higher in the tropics, but summer temperatures are lower than for much of the United States. Almost every place in the United States, except on high mountains or isolated peninsulas, has reached well over 100°F at times, and most places in the center of the country have reached 110° or 115°F. Even New York City on its island has hit 102°F, but Veracruz in tropical Atlantic Mexico has never exceeded 96°F; Colon, Panama Canal Zone, has never exceeded 95°F; and Belém, Brazil, virtually on the equator, has never exceeded 95°F. Places farther inland in tropical countries occasionally reach 100°F, but there is no record as high as 110°F for the tropical country of Brazil, although the United States has recorded 134°F. Furthermore, as we pointed out (p. 17), tropical areas do not typically have the cyclonic storms that provide alternating hot and cold and wet and dry spells in the temperate regions. The question this poses is, "Why does any animal or plant have a southern edge of its distribution between central United States and the equator?" Why, for instance, does the belted kingfisher, *Ceryle alcyon*, breed south on the mainland only to Texas, the Gulf Coast, and California? It gets cooler, rather than hotter, as we go south from this zone, so temperature is not the cause of this southern boundary. Some Mexican desert areas are doubtless too dry to provide the rivers in which kingfishers feed, but why don't these birds breed down the Atlantic coast of Mexico and reappear on the Pacific coast south of the deserts? We can rule out history in the form that says the species has never been down there, because belted kingfishers regularly winter in the tropics at least as far as Panama, where they are common and where winter temperatures are as high as summer. But every summer they come back to breed in the United States and Canada. If they can persist in Panama in the winter, there surely are streams, lakes, and ocean shores with fish suitable for them to feed on, so we can rule out quality of food as a factor. There are, however, two kingfishers, one smaller than the belted and one larger, that appear about where the belted stops. The small green kingfisher, *Chloroceryle americana*, appears in south Texas and at the Mexican border of Arizona; the very large ringed kingfisher, *Ceryle torquata*, has often been seen in Texas and breeds within a few hundred miles of both Texas and Arizona. These two clearly sandwich the belted kingfisher in the most uncomfortable kind of competitive squeeze (see p. 29). A wintering adult belted may be able to gather enough food for itself in their presence, but to gather food for a brood of hungry young is probably out of the question. In fact, farther south in Mexico two additional kingfishers appear, one very small and one about belted size, and by Nicaragua a fifth tropical species is present. The five tropical kingfishers are neatly arranged by size, and the farther south we go, as the climate is presumably less and less seasonal, the more species there appear to be. Even in Panama, however, two species of the five are rare, leaving the size sequence ringed, Amazon (*Chloroceryle amazona*), green, as the common species. Further evidence that the belted kingfisher competes with the others comes from looking at what happens in temperate South America as these kingfishers drop out. There are no belteds there, and the ringed kingfisher goes right down to Tierra del Fuego, with its perennial cold and its glaciers. It seems clear that in the absence of the belted it could do the same in the United States and Canada. Hence we tentatively infer that the southern edge of the belted kingfisher's range is due to competition in most places, although poor habitats—especially deserts—create local barriers; and we also infer that the northern edge of the ringed kingfisher's range is due to competition. Is this a frequent or infrequent situation? To answer the question we consider other species.

An examination of *A Field Guide to the Birds of Texas* (Peterson, 1963) reveals that 202 "land birds" (birds from pigeons on in the order that birds appear in the book) breed in that state. Comparing this list with *The Species of Middle American Birds* (Eisenmann, 1955) reveals that of these 202 Texas breeding species only 29 also breed in Panama. The remaining 173 have the southern edge of their ranges somewhere in between; although for some this must be because the habitat vanishes, for many it must be the effectiveness of tropical competitors that determines this southern edge. The kingfishers have such unique feeding habits that it was easy for us to suggest plausible competitors responsible for excluding the belted kingfisher as a breeding bird from the tropics. For other species it is much harder to suggest the culprits; in fact, usually there is no single species excluding one whose range suddenly ends; much more often a constellation of species must be present to exclude it. Our American robin (*Turdus migratorius*) breeds in the mountains south to the highlands of central Mexico. At least four new species of robins appear by the mountains of southern Mexico, as well as some others in the lowlands. They clearly are a sufficient reason for the American robin to drop out, but no single species can be blamed. All of these new robins are very much like ours in morphology and behavior; only the color and sometimes size are different. The new ones usually live at different elevations on the mountains. Panama has 564 breeding "land bird species" (pigeons on, as before) compared to the 202 in much larger Texas. Clearly most Texas species can expect to be greeted by a host of ecological counterparts at the southern edge of their ranges. Wholly new families of foliage gleaning birds, such as the antbirds (Formicariidae), appear in the tropics. No wonder so few of the American foliage gleaners make it to Panama, and those that do, like the yellow warbler (*Dendroica petechia*), are restricted in Panama by their competitors. The yellow warbler, for example, is restricted to mangrove swamps. We know yellow warblers are restricted by competitors because the island of Chitre, in the Pearl Archipelago, which has no antbirds or wrens, has yellow warblers as the commonest species in the forest. The nearby island of Rey, with the same predators but with three species of antbirds and a very common wren as well as other species not on Chitre, has its yellow warblers again confined mainly to the mangroves. There is, though, one astonishingly successful bird of eastern United States that actually retains its success in Panama. The red-eyed vireo (*Vireo olivaceus*), the most abundant breeding bird throughout our eastern deciduous forests, has a close relative (probably a race of the same species) usually called the yellow-green vireo (*Vireo flavoviridis*) that is almost equally successful in the tropics. In the tropics, however, the yellow-green vireo does not forage in the canopies of all broad-leaved forests; rather it is confined to second growth and is excluded from mature forests, probably by a combination of species.

In a few cases, southern edges of ranges are caused by a single extremely closely related species. Usually the two species form the semispecies of a superspecies. Thus among the tyrant flycatchers, two American species are replaced by almost exact tropical counterparts, the ash-throated flycatcher (*Myiarchus cinerascens*) by the pale-throated flycatcher (*M. nuttingi*) in Mexico, and the northern beardless tyrannulet (*Comptostoma imberbe*) (called beardless flycatcher in most American books) by the southern beardless tyrannulet (*C. obsoletum*) in Costa Rica. In interpreting such cases as these, we are at the mercy of the systematist. If he is sure there is no interbreeding between the semispecies, we are sure the range boundary is competitive; but if a large amount of interbreeding is discovered, the whole range border provides no evidence for competition.

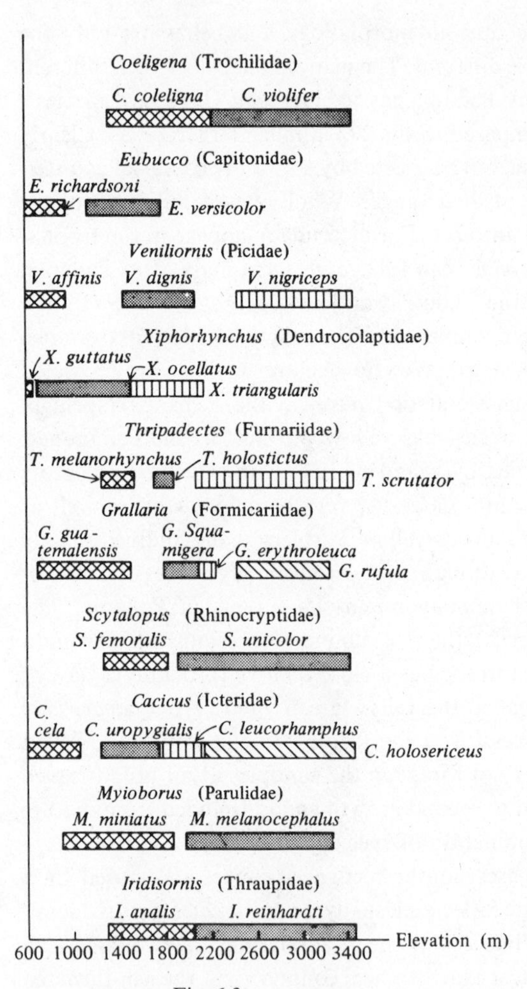

Fig. 6-3

Elevational replacement of congeneric species in ten families of birds on Peruvian mountains. Series of two, three, and four replacing species are represented. Note the apparent hiatuses between some pairs of species and the consistency with which the uppermost species possesses the broadest amplitude. (From Terborgh, 1971.)

In any case, the southern edges of species' ranges show many cases of nonclimatic range limitation, many of which are competitive.

Studies on mountains and islands. Mountain comparisons provide a major source of evidence for competitive range limitations but, again, chiefly from the tropics. We give one temperate zone example followed by several tropical ones.

On the higher New England mountains there are five species of thrush arranged more or less in horizontal layers. Veery (*Catharus fuscescens*) and wood thrush (*Hylocichla mustelina*) are mainly confined to the deciduous forest and are found up to about 1700 ft. The hermit thrush (*C. guttata*) appears in patches of pines at about 1000 ft, or occasionally lower, and occurs up to over 3000 ft wherever conifers are found with fairly barren understory. Beginning at about 1800 ft in wet spruce and hardwood, the Swainson's thrush (*C. ustulata*) appears and continues right up to the stunted firs at timberline. Finally, the gray-cheeked thrush (*C. minimus*) occurs in the stunted and windblown firs and spruces near the timberline. In the Great Smoky mountains as in New England there are both dry and moist fir and spruce forests of the kind hermit and Swainson's thrushes should like; but those species are absent, and in their absence the veery extends into the fir forests high on the mountains. We tentatively conclude that the veery is missing from this zone in New England because of competition from hermit and Swainson's thrushes.

Terborgh (1971) has given excellent examples from Peru, which we illustrate in Figs. 6-3 and 6-4. The first shows congeners whose altitudinal ranges abut or leave gaps, and the second shows a genus whose species occupy overlapping elevations. These illustrate two points that Terborgh proves statistically using data from nearly 5000 birds captured in mist nets. First, species higher on the mountains occupy a wider range of elevations. This is correlated with a reduction in numbers of high-elevation species and is apparently an island effect. Island species do occupy a wider range of habitats. The second point is that, at a given elevation, the members of a three-species replacement occupy a smaller range than the members of a two-species replacement, and these occupy a smaller range than a single species. This would be easy to predict on the basis of competition. To simplify, if a genus of three species occupies 2400 vertical meters with nonoverlapping distributions, each will occupy an average of 800 m; if a two-species genus subdivides the 2400 m, each will occupy an

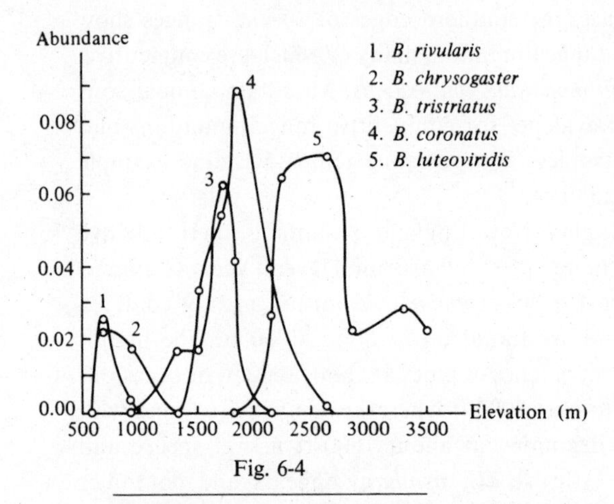

Fig. 6-4

Population density curves for species in the genus *Basileuterus*, Parulidae (warblers), in Peru. Species 1 and 2 overlap but are of different size, occupy different habitats, and forage at different levels in the vegetation. Species 2, 3, and 5 replace each other without overlap, while species 4, which differs in size from both 3 and 5, reaches maximum abundance in the replacement zone. (From Terborgh, 1971.)

average of 1200 vertical meters. This kind of pattern is a consequence of competition and cannot be accounted for by any hypothesis of independent distribution of species. Terborgh found that the bottom members of the sequences (those with the smallest vertical ranges by his first conclusion) had the following average vertical ranges: 64 cases of a single species in the genus, 833 m; 19 cases of two-species sequences, 613 m; 17 cases of three-species sequences, 440 m; 14 cases of four-species sequences, 363 m. He most reasonably interpreted this as strong evidence for competition. (Terborgh's examples do not conform to the hypothetical cases of two- and three-species genera mentioned above, because in the first place his combined species do not necessarily occupy the whole mountainside, and in the second place the congeners haven't subdivided the elevation equally.)

Diamond (1970a) has made similar studies over several years in the mountains of New Guinea and on its offshore islands. He, too, has found that competition plays a major role in controlling bird distributions. For instance, three cases are illustrated in Fig. 6-5. We see immediately that where a mountain range is missing a species, its congeners

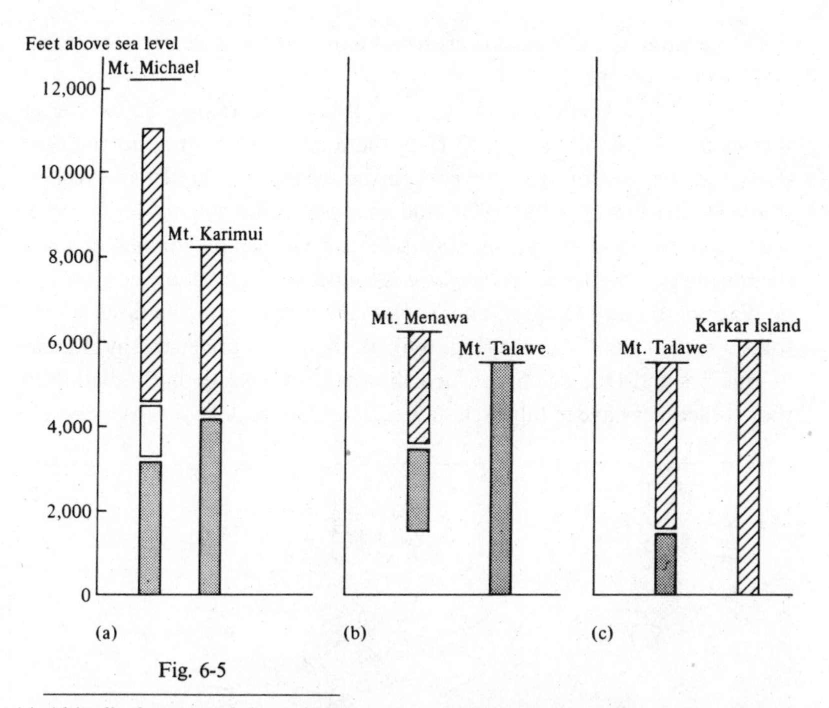

Fig. 6-5

(a) Altitudinal ranges of three similar congeneric flower peckers on two New Guinea mountains: *Melanocharis nigra*, solid bars; *M. longicauda*, open bar; *M. versteri*, shaded bars. The elevation of the summit of the mountain is indicated by the horizontal line. On large mountains such as Mt. Michael, all three species are present and have mutually exclusive altitudinal ranges (0–3,200, 3,200–4,550, 4,550–11,000 ft). On the smaller and more isolated Mt. Karimui, *M. longicauda* failed to establish itself, and the remaining two species expanded their ranges to fill the gap: *M. nigra* at 0–4,200 ft, *M. versteri* from 4,200 ft to the summit at 8,200 ft. (b) Altitudinal ranges of two congeneric white-eyes on mountains of two Pacific islands: *Zosterops atrifrons*, solid bars; *Z. fuscicapilla*, shaded bars. On New Guinea (e.g., Mt. Menawa), both species are present and have mutually exclusive altitudinal ranges, *Z. atrifrons* at 1,500–3,500 ft and *Z. fuscicapilla* at 3,500–6,000 ft. Only *Z. atrifrons* succeeded in colonizing New Britain (Mt. Talawe), where it occurs abundantly up to the summit at 5,500 ft, through the altitudinal range utilized by *Z. fuscicapilla* on New Guinea and thus 2,000 ft higher than in the presence of its competitor, as well as extending in low numbers down to sea level. (c) Altitudinal ranges of two congeneric lorikeets on mountains of two Pacific islands: *Charmosyna placentis*, solid bars; *C. rubrigularis*, shaded bars. Both species are present on New Britain (Mt. Talawe) and have mutually exclusive altitudinal ranges, *C. placentis* at 0–1,500 ft and *C. rubrigularis* at 1,500–5,500 ft. Only *C. rubrigularis* succeeded in colonizing Karkar Island, where it extends down to sea level, having expanded into the whole range of its absent competitor. (From Diamond, 1970a.)

tend to enlarge their distributions to take over. This is clearly most easily explained by competition.

Studies of superspecies. Where the ranges of two semispecies meet with very slight overlap, there is an excellent opportunity to study the competition that prevents further range overlap. In an excellent study of this kind, Smith (1970, and manuscript in preparation) studied the two squirrels of the genus *Tamiasciurus* that occupy conifer forests in the northwest. *Tamiasciurus douglasii* is found mainly west of the Cascades in Washington and Oregon, and *T. hudsonicus* mainly in the drier conifer forests east of the Cascades (Fig. 6-6). Both species prefer the light cones of true firs and Douglas firs, which they can open easily, but in bad years they sometimes have to fall back on less desirable food. In the drier country

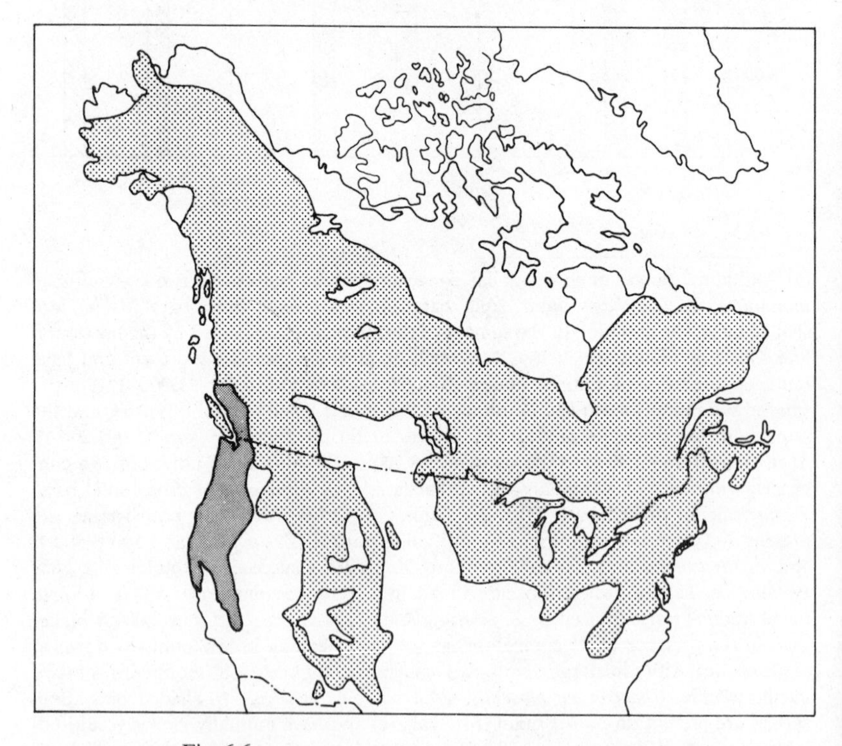

Fig. 6-6

Present geographic distribution of *Tamiasciurus hudsonicus*, the red squirrel (stippled areas), and *T. douglasii*, the Douglas squirrel (shaded areas). (From various sources.)

east of the Cascades, the cones of lodgepole pine (*Pinus contorta*) are the main food they turn to when fir cones are gone. Here the pine cones are of a closed type requiring strong jaws to open. The somewhat heavier *T. hudsonicus* has stronger jaws (Fig. 6-7), and Smith showed that it could eat these cones more effectively. Hence it has the advantage east of the Cascades in the dry country. West of the Cascades there is no abundance of lodgepole pine cones to fall back on, and in a poor year for cones the surviving squirrels are those that are light and agile enough to reach the cones on the ends of very fine hemlock branches and the seeds of the alder, birch, and maple trees. The lighter *T. douglasii* has a double advantage here. Not only can it reach the tips of the smaller branches, but it also requires fewer cones because of its smaller size. In addition, it has not wasted energy developing large jaw musculature. Finally, Smith concludes that each gets a slight additional advantage in its own range by protective coloration; the darker *douglasii* is less conspicuous in the dark damp coastal forests, and the lighter *hudsonicus* is less conspicuous in the opener and drier inland forests. Smith also discusses effects of comparative behavior and litter size on the semispecies' coexistence.

In another study, Brown (1971) has studied the two chipmunks *Eutamias dorsalis* and *E. umbrinus*, whose distributions barely

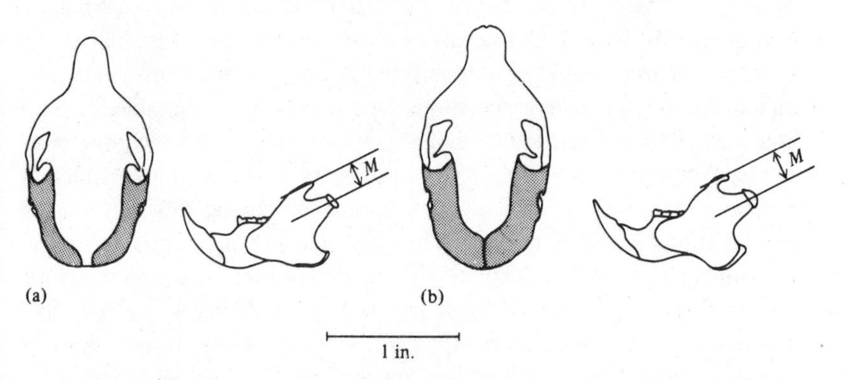

Fig. 6-7

Dorsal view of the skull and lateral view of the lower jaw of a Douglas squirrel (a) and a red squirrel (b). The shaded area of the skull is the surface of attachment of the temporal muscles. Skull (b) has a sagittal crest and skull (a) does not. *M* is the distance of the moment arm of the force applied by the temporal muscles. Its greater distance in (b) allows a greater force to be applied by the teeth. (From Smith, 1970.)

overlap on isolated western mountain ranges. *Dorsalis* excludes *umbrinus* from the sparse piñon-juniper forests of lower elevations by its skill in fighting. Where the forests become denser and connected pathways through the canopy are possible, the advantage shifts to the more arboreal *umbrinus*. In these habitats *dorsalis* seems to waste its time and energy attempting to fight.

The Role of Habitat

In the temperate zone at least, more species seem to have ranges limited by habitat than by any other factor. On the other hand, Terborgh (1971) showed that most Peruvian birds did not have their altitudinal zones bordered by the distinct changes in habitat. But even where this factor does limit a species' distribution, we can legitimately ask why the species has not adapted to a wider range of habitats. To this question we can give two main answers, which we illustrate with examples.

1. On page 131 we considered the delayed germination of seeds of desert annual plants. If they have all seeds germinating, in a year of inadequate rain all plants will die and the species will vanish. Thus we can see why an annual plant that has all its seeds germinate cannot live in deserts. Conversely, a desert annual with but half of its seeds germinating would be outcompeted in a predictably moist climate by an annual with all seeds germinating. Neither strategy is suitable for both climates, and this is an example of a general situation in which one morphology or physiology simply does not work in both of two different habitats. Plants often counter this dilemma by having different genetic combinations in different areas. Thus it would not be surprising to find an annual plant in a moist climate with a genotype for all seeds germinating, and to find the genotype for half the seeds germinating where only half the years are suitable. But even a plant species cannot always adjust to abrupt habitat changes. Water lilies are confined to water, and cacti to the land. Looking at animal examples, we find that ducks, whales, and seals have committed themselves in a way that precludes foraging for insects in the canopies of trees! Where their habitats give way to forest, they disappear.

2. Competition or predation may restrict the habitat of a species even though it is capable of occupying a broader one in the absence of competitors. We saw (p. 135) that yellow warblers (*Dendroica petechia*) were confined to mangroves over most of the tropics, but on a few islands where competitors were few they occupied all the forests. Hence their restriction to mangrove habitat is most likely a result of competition. Diamond observed habitat expansion among 12 of 52 New Guinea bird species on the island of Karkar, from which he could infer that about one-quarter of the New Guinea species have competitive limits to their habitats.

Having given examples of each kind of habitat limitation we proceed with a few more examples that are of considerable interest, though all need further study. Joseph Grinnell, who knew a vast amount about biogeography half a century ago, provided several interesting examples. On studying the California thrasher (*Toxostoma redivivum*) (Grinnell, 1917), he concluded that it occupied chaparral and not more open oak woodland because of predators. He reasoned that the food supply was equally good in the oaks but that the thrasher would be more vulnerable to hawks there. Its absence from the Arizona chaparral is most readily explained by competition, however, for in Arizona the very similar Crissal's thrasher (*T. dorsale*) occupies the manzanita chaparral.

Grinnell and Orr (1934) and Grinnell (1914) compared the distributions of large species with smaller relatives. Grinnell and Orr pointed out that the very large mouse, *Peromyscus californicus*, is much more restricted in habitat than are the smaller *Peromyscus* species. In their picturesque language, "... the small-gauged mouse will find shelter suited to its requirements for existence more continuously and farther than the larger-gauged mouse . . . (The larger) is dependent upon the availability, in sufficient frequency for a persisting population of its own kind, of larger-sized crevices, spaces, or holes." Grinnell had earlier come to the same conclusion on comparing the habitats of the large kangaroo rat, *Dipodomys deserti*, with small *D. merriami*. *Merriami* was found in nearly every desert habitat, while the large *deserti* was confined to deep sand of over an acre in extent, where it could dig its large nest holes. Grinnell concluded that larger species usually have more restricted habitats than their smaller relatives.

Statistical attributes of habitat selection can also be exhibited. We can consider three kinds of small rodents: the microtines, the genus *Peromyscus*, and the genus *Perognathus*. The microtines, including the common meadow mice, eat grass and other moist vegetation. Those that eat mainly grass have continuously growing molar teeth to counteract the wear caused by silica in the grass. The mice of the genus *Perognathus*, at the other extreme, have large cheek pouches for carrying seeds and can survive without any source of water or moist vegetation. They store seeds in their burrows and may hibernate in bad seasons. The *Peromyscus* mice are intermediate. They will eat both seeds and succulent vegetation (but probably not grass except its seeds), they store some food but do not have special cheek pouches for the purpose, and they do not hibernate. The adaptations of microtines clearly limit them to areas of continuously present grass or herbaceous vegetation; they are excluded from deserts. *Peromyscus* can live in either deserts, grasslands, or forest but is less abundant in most grassland than microtines and less abundant in deserts than *Perognathus*. *Perognathus* is virtually confined to deserts and other dry habitats.

The number of examples in this section could be multiplied indefinitely. Special ant species are confined to the trees *Cecropia* and an *Acacia* with which they live mutualistically. The ants viciously attack an animal molesting the tree and may prune competing vegetation, and the trees provide shelter and nourishment for the ants. Hole nesting birds are confined to the trees or banks that provide the holes, butterflies are confined to the vegetation types that contain their larval food plants, and so on. In each case we may legitimately ask why there is just the particular habitat limitation, and whether competition and predation play a role.

The Fluctuating Boundaries of Species Distributions

Although we have presented discussions of the boundaries of species' distributions as if these boundaries were immutably fixed, they actually fluctuate greatly. Of the controls we have discussed—climate, competitors, predators, habitat via adaptations, habitat via competitors and predators—perhaps only pure habitat by way of adaptations is not subject to yearly fluctuations, and even habitats fluctuate slowly, as we shall study in Chapter 9. Certainly climates differ from year to year, and as the climate varies so may the food supplies. This causes a change in the region of coexistence of two competitors, a region of subtle balance in the abundances of foods. In short, we expect the boundaries of a species distribution to fluctuate.

Part of the difficulty in documenting fluctuating range boundaries is that these boundaries can be very tenuous (see Fig. 6-8). Locating the northernmost individual of a population is a very difficult

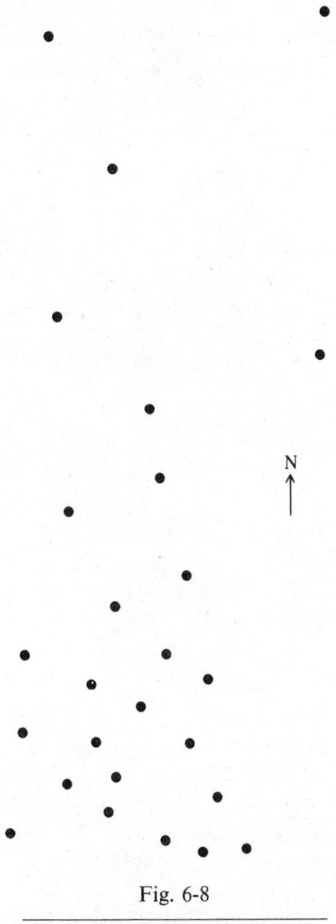

N

Fig. 6-8

If individuals of a species are located as shown in this figure, what is the edge of the species' range? See the text for a discussion.

exercise and is of doubtful value in any case. Should we say the northern edge of a range is where the northernmost individual is found? It would really be much better to say it is where the northernmost colony of reproducing individuals is located. And this colony may go extinct randomly, or with change in climate or the balance of competition. We could also, arbitrarily but usefully, define the northern edge of the range as the point where the population density falls to, say, one-twentieth of its density in the middle of the range or one-twentieth of its maximum density. This kind of

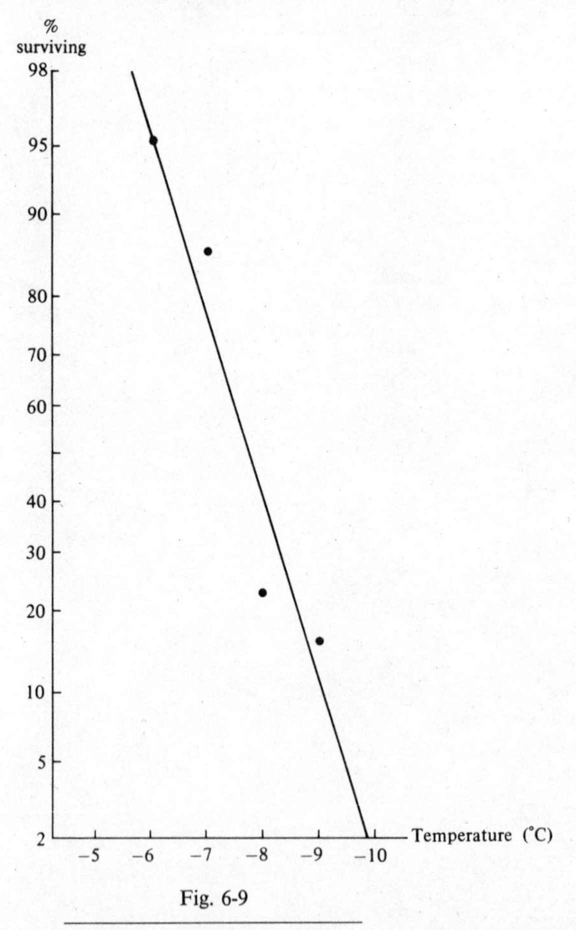

Fig. 6-9

Survivorship of *Drosophila nigrospiracula* (Tucson Mountains, Arizona) in the supercooled state. (From Lowe, Heed, and Halpern, 1967.)

definition can occasionally lead us into difficulties but has the great advantage that it sets us to counting individuals where there are still enough to count. By such a definition, too, there will be fluctuations in the species range, although chance will play less part in them.

The record cold weather of January 1971 killed back many plants near Tucson, Arizona. As we have seen (p. 127), the saguaro cactus lives only where it thaws every day. But in January 1971, as in January 1962, which was also a severe winter, many places that normally thaw every day had 36 hours of continuous frost. Not only did saguaros die in numbers, but the range of the desert ironwood (*Olneya tesota*) was severely cut back. The ironwood was formerly abundant eastward to the east slopes of the Tucson Mountains (just west of Tucson). The individuals on this east slope died, or at least suffered severe damage, during the 1971 frost. The very nature of the mortality caused by such an event is vague and is usually described by a "dosage mortality curve," just as is done in drug studies (Fig. 6-9). Even in a very light frost some individuals die, and even in a very severe frost a few individuals survive. Figure 6-10 shows the saguaro mortality at one location due to the 1962 freeze.

Man has presumably been responsible for many range changes. Not only have large predators been killed back so that there are

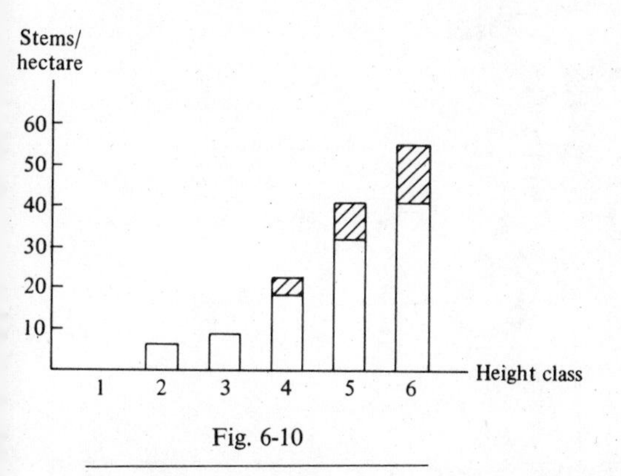

Fig. 6-10

Population histograms for saguaro in the area of Tucson, Arizona. The heights of the bars indicate numbers of individuals per hectare (2.5 acres). Height classes are as follows: (1) up to 1 ft (0.3 m), (2) 2–6 ft (0.6–1.8 m), (3) 7–12 ft (2.1–3.6 m), (4) 13–18 ft (3.9–5.5 m), (5) 19–24 ft (5.7–7.3 m), and (6) over 24 ft. Shaded portions of bars represent recently dead individuals, most of them killed in the freeze of January 1962. (From Neiring, Whittaker, and Lowe, 1963, after Turner and Hastings.)

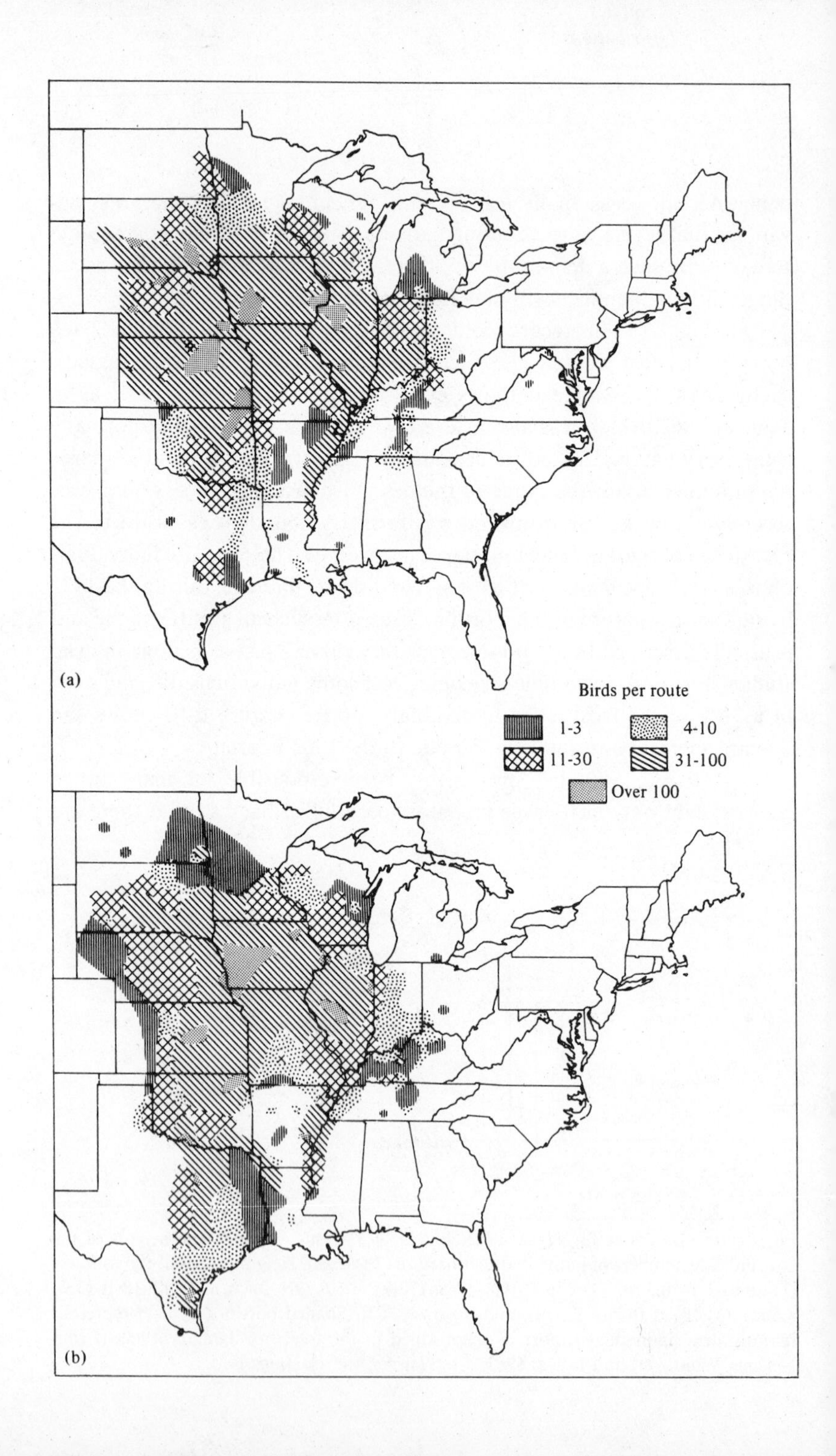
(a)
Birds per route
1-3
4-10
11-30
31-100
Over 100
(b)

now only a handful of wolves left in the United States, but inadvertent effects have been remarkable. For example, the dickcissel (*Spiza americana*) is now considered a bird of the tall grass prairie, but its breeding range not only fluctuates greatly from year to year (Fig. 6-11) (Robbins and Van Velzen, 1969) but it has had even greater fluctuations in history. It used to be a common breeder in fields along the Atlantic coast, then dwindled and disappeared from this area between 1860 and 1880. S. Fretwell (pers. comm.) has suggested that its breeding range may expand or contract according to the adequacy of the winter food supply; he suggests the decline in the 1860s was due to a change from rice growing to cattle raising in the South American wintering range.

Man has presumably also been directly responsible for the northward spread of many species that utilize bird feeders. Thus, cardinals (*Cardinalis cardinalis*) and tufted titmice (*Parus bicolor*) have spread northward and, possibly for the same reason, the evening grosbeaks (*Coccothraustes vespertinus*) have spread south and east. All are heavy frequenters of the vast numbers of bird feeders that now cover the country.

The Structure of a Species Population

The north, south, east, and west boundaries of a species' range tell us very little about what is happening inside, which is the topic of this short section. It is short because knowledge is so meager.

It was a favorite idea of Grinnell (1943) that reproduction is fastest in certain areas where the species is best adapted to its environment. In these same regions mortality is likely to be lowest. Here there will be a large net increase in individuals, which will consequently migrate away to less favorable areas. Hence, Grinnell believed, there will be centrifugal flow of individuals radiating outward from centers of best adaptation. Viewed from a perspective of well over half a century, there

Fig. 6-11

Dickcissel breeding ranges in 1967 (a) and 1968 (b). (From Robbins and Van Velzen, 1969.)

seems only one doubtful point about Grinnell's reasoning. Why would any individual ever migrate to a less favorable area? Why not stay put if it is better at home?

We proceed to answer these questions. The per capita birth rate usually falls with increasing population density, and the per capita mortality rate certainly rises with increasing population density. Hence, for each region, there will be a density at which mortality will just balance new births and the population will be at equilibrium (Fig. 6-12). We first ask whether it would ever be desirable for an individual from one such equilibrium population to migrate to another. Probably, the answer is no. In each equilibrium population an average individual just replaces itself with one offspring (i.e., two surviving offspring per pair). Thus the individual contemplating migration would not be entering a population where it would leave more offspring. Rather, in the unfamiliar territory it would tend to leave fewer than average and thus do itself harm by the change. A more substantial migration maintaining the equilibria would do

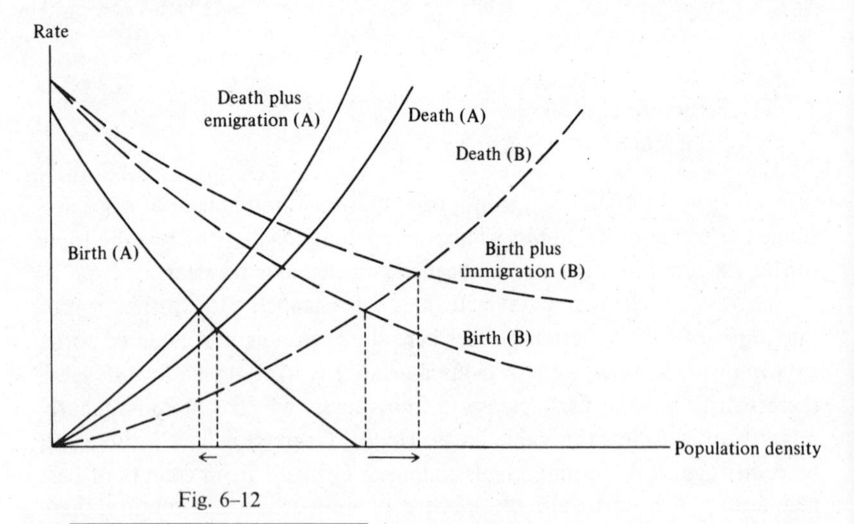

Fig. 6–12

Birth and death rates as functions of population density in habitats A and B. When emigration is added to the death rate in habitat A, its equilibrium population shifts to the left as indicated by the arrow. When immigration is added to the birth rate of habitat B, its equilibrium density shifts to the right as marked by the second arrow. At the new equilibria, birth rate is greater than death rate in habitat A, and birth rate is less than death rate in habitat B.

even more harm. Suppose there were a regular flow from population A to population B. Then A will have a reduced equilibrium population (its mortality curve will be augmented by the rate of emigration) and the B population will have a higher equilibrium as its birth rate will be augmented by the rate of immigration. This means that in population A mortality is less than birth rate alone (since mortality *plus* emigration equaled birth rate) and a stay-at-home could expect to leave more than one surviving offspring. Similarly, in population B, birth rate alone would be less than mortality at equilibrium, because birth rate plus immigration would equal mortality. Hence an immigrant could expect to leave fewer than one surviving offspring. Migration under these equilibrium circumstances would certainly be selected against.

Is Grinnell's view for once wrong then? Are there no circumstances when migration to a new area would be favored? It is a truism that migration would be beneficial if the organism were better off afterward, but this is not difficult to arrange. If the population in area A has just gone extinct even though the area is suitable for the species, then migrants from nearby area B will certainly be favored, for they may leave far more descendants than they would have had they stayed at home. More technically stated, they will migrate if their "reproductive value" would be greater. This is essentially why a bird wintering in the tropics may migrate to Canada to breed; it is recolonizing an area left empty during the winter but now in the flush of new growth. The species reproduces so vigorously in Canada that in spite of the hazards of the migration, it fares better than it would have by staying in the tropics. Of course, the more species migrate to Canada, the less desirable it becomes, the more desirable the tropics become, and eventually an equilibrium is reached such that again the migrant would be worse off than if it had stayed at home. Thus there are circumstances when a migrant population is better off than one that stays at home, but this beneficial kind of migration might better be called *pulses* of migration because the individuals return at another season. Pulses of migration are often beneficial. In good years populations enter marginal areas, only to leave them in bad years. The poplars, pines, spruces, and birches that sprout up in abandoned fields are later replaced by trees that cast a shade so deep that the first species cannot compete. The poplars and the others then, the species early in the succession, are obligatory pulse migrants. They must perennially colonize newly vacated spots and suffer replacement where they are. Salisbury (1942) counted and weighed the seeds on herbarium specimens of many species of plants and found, as we would expect, that the early succession species have turned their reproductive effort to very many light, often windblown, seeds. The trees later in succession, such as the beeches and oaks, produce instead a few very large seeds. These do not disperse so well but give a young seedling in a competitive situation a good start.

Returning to Grinnell's proposition, can we ever find a situation where a unidirectional and continuous migration is beneficial? The answer is unknown.

Figure 6-12 did show that populations will be denser at equilibrium in some habitats than in others, even if we could not demonstrate any reason for a population flow from the dense to the sparse. Thus some habitats hold a dense population, some hold a sparse population, and some hold no individuals at all. Can we tell which habitats a species will occupy? Fretwell (1972) devised a striking way of portraying the answer (Fig. 6-13). The excess of per capita birth rate over per capita death rate (which we have often called r) is plotted against population density as a separate curve for each habitat. When the density in habitat 1 reaches the level D_1, individuals colonizing habitat 2 will be just as fit there as in habitat 1 and the population will begin to occupy the second habitat. When the population density reaches level D_2, individuals in habitats 1 and 2 are no longer leaving more offspring than they could in habitat 3, and its colonization begins. But when the population density reaches D_3, the death rate just matches the birth rate so no further population increase can occur. The population will normally never become dense enough for habitat 4 to be worth colonizing. At the equilibrium, the r value in all habitats will be zero, so the population densities in the three occupied habitats are given by the lower intercepts of the three curves in the figure. Habitat 1 will have density D_3, habitat 2 will have D_5, and habitat 3, D_6. Fretwell points out some cases where these densities appear at first sight not to hold. Field sparrows (*Spizella pusilla*) that he studied in North Carolina did not reach just the breeding density at which their breeding was successful but instead bred in greater densities, at which breeding success was less. But during the winter all individuals had to feed in the field that had had the high breeding density, and since the individuals that had bred there had practiced in this habitat they appeared to survive better in the winter.

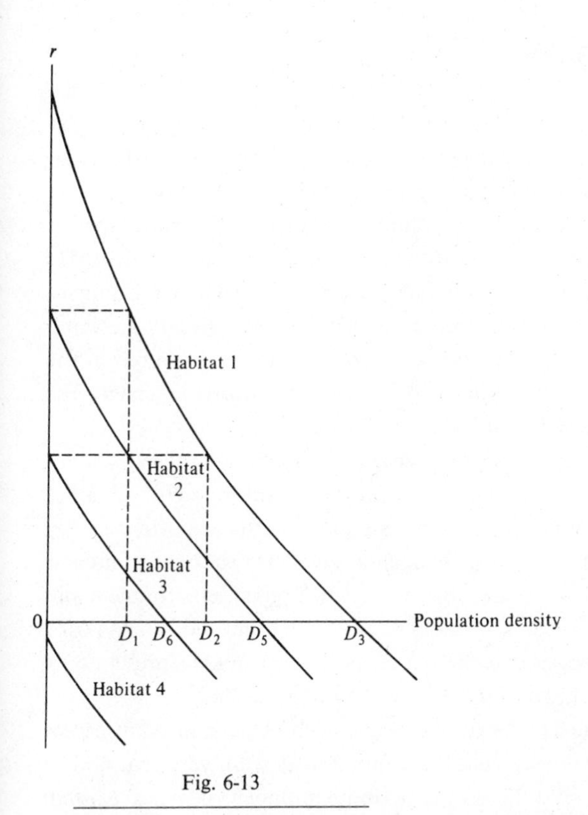

Fig. 6-13

r as a function of population density for four habitats. See the text for details.

Since breeding and mortality occur largely in different seasons for birds and since birds often migrate to different habitats in different seasons, Fig. 6-13 is virtually impossible to apply in practice. If one could assume that all southward-bound migrants were equally likely to survive the winter (untrue in the field sparrow), we could replot the figure with birth rate in place of r. Then the population densities would no longer equilibrate along the abscissa but rather along some undetermined horizontal line; and the figure would tell us in which habitats the species should breed.

The greatest difficulty in testing the models of this section is only lack of ambition. It would be hard work to find the curves of the figures for different species in different habitats, but it is not beyond present techniques.

Communities of Species

Perhaps the most important question we can ask about communities is, "Does the environment dictate the structure of the community, or are the species a fairly random assemblage?" Then we can ask, "Are the boundaries of these communities sharp, with many species dropping out synchronously, or do the species drop out independently?" The second question is clearly different from the first: The environment could exert a profound control over the morphologies, physiologies, and numbers of species, and the first question would thus be answered in the affirmative; yet as the environment varied continuously in space, the different species might drop out independently.

Perhaps the most extreme view of communities was advocated by Andrewartha and Birch (1954), in their book on ecology. Since they believed species interactions were generally of little importance, they assumed each species lived its life independently. They thought of a community as a collection of species all adapted to a given environment but living independent lives. They would seldom expect and did not look for either community or species diversity patterns. These would be present only if species interacted, a proposition which they had rejected.

At the other extreme, some early American plant ecologists like Frederick Clements believed communities were very real and of fairly rigid structure, and that only a finite number of types existed. Initially intermediate communities would converge to one or another of these "climax" types. In fact, many ecologists, at that time and even now, believed that communities were superorganisms endowed with properties that could not be understood by an examination of their components. A great deal has been written about "the whole being greater than the sum of its parts." The trouble with this is that "greater" and "sum" are not appropriate words for analyzing wholes and parts. Most scientists believe that the properties of the whole are a consequence of the behavior and interactions of the components. This is not to say that the way to understand the whole is always to begin with the parts. We may reveal patterns in the whole that are not evident at all in its separate parts. Species diversity, for example, is a community property and is not a property of the individual component species. It can be understood as a consequence of the interactions of these species, as we shall see in the next chapter, but its patterns were discovered and explained by people aware of communities; ecologists primarily interested in the separate species have never made any progress in unravelling community patterns.

Environmental Control of Community Structure

That the forms of species in a community are under some environmental control is obvious to everyone. The mammals living in the ocean are obviously different from those on land, and the plants in the deserts are clearly different from those of wet warm lands. But it is not so immediately clear how subtle the environmental control is and with what minute precision it acts.

The first clues on this issue come from convergence (Fig. 6-14). Animals and plants of differing phylogenetic ancestries but

Fig. 6-14

(a) Typical blue tit *Parus caeruleus caeruleus* of broad-leaved forest and (b) thin-beaked Canarian form *P. c. teneriffae* of pine forest; (c) typical coal tit *Parus ater ater* of conifer forest and (d) thick-beaked Persian form *P. a. phaeonotus* of oak woods. (From Lack, 1971, by Robert Gillmor, adapted from David Snow.)

occupying similar environments come to look very much alike. The jerboas of African deserts and the kangaroo rats of American deserts are very similar in form but not in ancestry; they must have converged under the influence of similar environments. The American wood mice of the genus *Peromyscus* are extremely similar in appearance to the European wood mice of the genus *Apodemus* belonging to a different family. So similar are they, in fact, that a person familiar with *Peromyscus* would pass over an *Apodemus* as just another *Peromyscus*. The cactus typical of American deserts closely resemble some *Euphorbias* of Old World deserts. In each case, the closeness of the convergence indicates how precisely the environment tailors the morphologies of its occupants. But these are individual species; do the community properties converge, too?

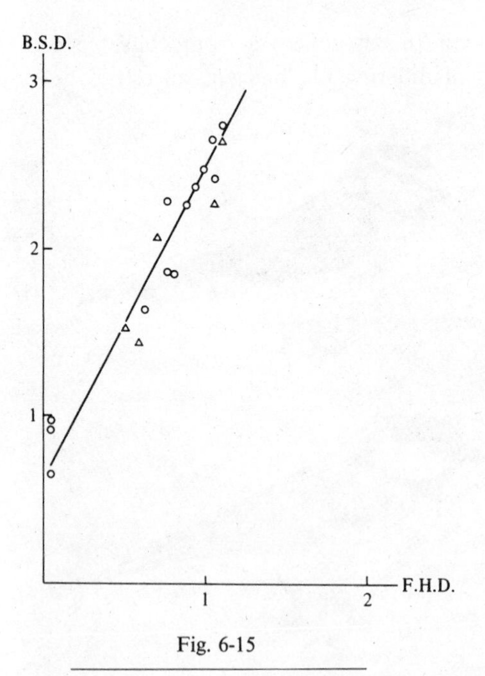

Fig. 6-15

Comparison of the relation between bird species diversity (B.S.D.) and foliage height diversity (F.H.D.) in Australia and North America. The circles are North American points and the triangles are Australian points. The regression line is based on the North American data. (From Recher, 1969.)

The numbers of species are similar in similar habitats around the world. Recher (1969) studied Australian birds and found (Fig. 6-15) that their diversities exactly paralleled those of American birds in comparable habitats. That the Australian line parallels the American one is a clear case of community convergence, although the exact coincidence may be fortuitous.

The next level of demonstration that species in Old World communities have counterparts in the corresponding New World communities is revealed in an example provided by Lack (1971). He has extensively studied tits (family Paridae) that form a large part of the foliage gleaning bird community in England, and he has assigned to each its role in gathering food. Some eat large foods but have to perch on stiff twigs; others are lighter and can perch nearly anywhere but eat smaller food. In every case the bill size and shape is nearly perfectly correlated with the foraging place and food type. This remarkable fact is made clear by Lack's comparison of tits in the Old and New Worlds (Fig. 6-16). Each European species has an American counterpart that not only is of the same size and bill shape but also feeds in the same sort of situation. However, as many as six of these species live in the same community in England, whereas more than two (a large and a small) seldom coexist in the United States. The diffuse competition of wood warblers, vireos, and tanagers doubtless makes the way of life of these tits more precarious than in England, so they form a lesser part of the community.

The most detailed study that has been made along these lines was undertaken by Cody (1968), who compared the grassland bird communities in various parts of North America and Chile. He measured the degree of species separation by vertical foraging height, VHS (vertical habitat separation), the degree they were separated horizontally by microhabitat, HHS, and the degree of separation by feeding behavior, FS, for each community and found that the sum of the separations was essentially constant for all habitats in North or South America (Fig. 6-17). In fact, fields of similar structure in the two continents had counterpart bird species, although of different ancestries. Thus the American bushtit (*Psaltriparus minimus*) has a Chilean counterpart in a bird of flycatcher ancestry, and the California thrasher (*Toxostoma redivivum*) has a Chilean counterpart in the ovenbird family.

Tufted tit
Great tit
Bridled tit
Blue tit
Black-capped chickadee
Willow tit
Carolina chickadee
Marsh tit
Mountain chickadee
Crested tit
Chestnut-backed chickadee
Coal tit

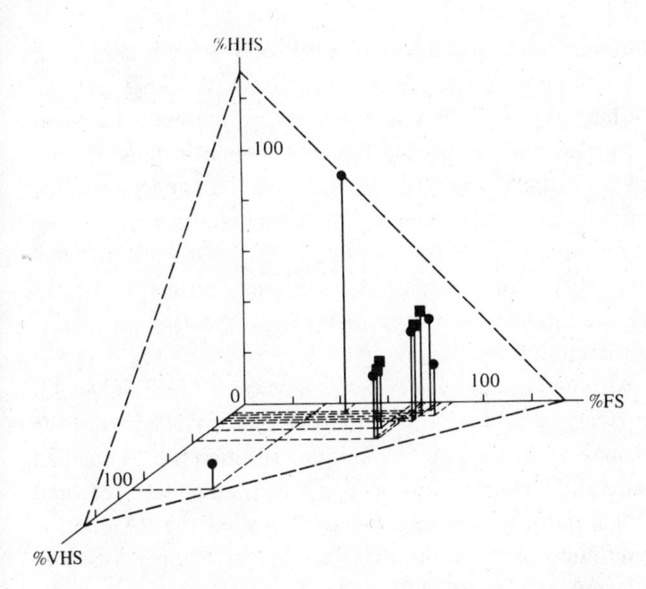

Fig. 6-17

Three-dimensional graph of ten grassland communities showing how species differ ecologically in these communities. The points lie approximately on a plane whose intercepts on the three axes do not differ significantly. Circles are North American points and squares are South American points. See the text for details. (From Cody, 1968.)

Distribution of Vegetation Types

Since many plants have boundaries of their distribution determined climatically, we expect large changes in climate to produce large changes in the particular combination of species present. At some point we would find so few of the original plant species remaining that we would say we were in a new vegetation type. Thus, for instance, a New

Fig. 6-16

Six North American tits (left) and their European counterparts (right). (From Lack, 1971, by Robert Gillmor.)

England mountain may have the "northern hardwood" forest, composed mostly of maples and birches and beech, at its base. At 3000 feet the adiabatic cooling has taken its toll and "northern coniferous" forest of spruce and fir with a few birches left has replaced the hardwoods. Higher yet, the cold and winds have taken further toll and the conifers become stunted and eventually disappear, replaced by alpine plants, often of the same species as are found beyond the tree line in the arctic. We presume many of these are relicts from the days when glaciers pushed arctic climates past these mountains; on the retreat of the glaciers the plants could remain on the mountains.

Although the point of view is now out of fashion, ecologists half a century ago would have said that we had travelled from the "Transition life zone" at the base of the mountain through the "Canadian life zone" and into the "Hudsonian life zone" in the stunted trees and adjacent tundra. The pure tundra might even be called the "Arctic life zone." Moving south to southern Florida one would add "Upper Austral," "Lower Austral," and "Tropical" life zones. These zones were first defined by Merriam (1890) through a trip up to the San Francisco Peaks in Arizona, where a similar sequence of vegetation types corresponded to the same sequence of zones. In Southern Arizona the Sonoran Desert would be Lower Austral (often called Lower Sonoran in the west); oaks and piñon-juniper country would be Upper Austral; the ponderosa pines higher on the mountains would be Transition; Douglas fir, Canadian zone at the top of these mountains, and if there were Englemann spruce and alpine fir Hudsonian zone would be reached.

The attraction of these terms lay in some overall similarities. The bird watcher familiar with the New England "Canadian zone" forest would find the same bird families and about one-third of the same species on a visit to the Arizona "Canadian zone" forests. (Hermit thrushes, myrtle-Audubon's warbler, pine siskins, hairy woodpeckers, solitary vireos, and golden-crowned kinglets would be among the familiar species.) Merriam actually defined these life zones in terms of temperature —winter and summer combined in a particular way. But he quickly added that once the zones were defined, it was convenient to use the plants themselves as indicators.

Life zones were subject to drastic criticism in the 1920s and 1930s largely on two counts. First, an error in Merriam's actual calculations was detected. But this error was applied equally to all zones, so that it did not alter their relative positions, and this is all that matters. Hence we can dismiss the first criticism. They were further criticized on the grounds that moisture was ignored. The Lower Austral of Arizona deserts was very unlike that of southeastern cypress swamps, and in this instance they contained no bird species in common except those like vultures that are not fussy. The critics were properly constructive, not confining themselves to complaining about life zones but proposing an alternative, "biomes." But biomes were described in terms of the characteristic plants and were really very different. Life zones were, at least in conception, an attempt to use climate to predict vegetation whereas biomes were no more than descriptions of existing vegetation. With life zones one could hope to visit an unexplored continent and predict its vegetation on the basis of its climate.

More recently Holdridge (1967) has attempted to revive the life zone approach, and his scheme at least has the merit of using both temperature and moisture in a general classification of the world's vegetation. Having seen the complexity of the climatic control of plant species, we cannot expect any such simple scheme to predict all changes accurately, but it can give us a good general idea (see Fig 6-18).

A critical question remains: Do different plant species change synchronously, or does each have independent distribution? If they change synchronously, vegetation types are more than a mere convenience; they are real and hence necessary as a subject of study. Whittaker (1969) has spent much of his life investigating this and has shown fairly convincingly for mountains in the United States that plants appear and disappear independently as we go up a mountain (Fig. 6-19). Holdridge might dispute this, for the tropics at least, where he believes plants change synchronously. However, no one has carried out in the tropics a study like those of Whittaker and we must await such a study before we can pass final judgment on whether life zones are real in nature or whether they are the scientist's convenient but arbitrary classifications.

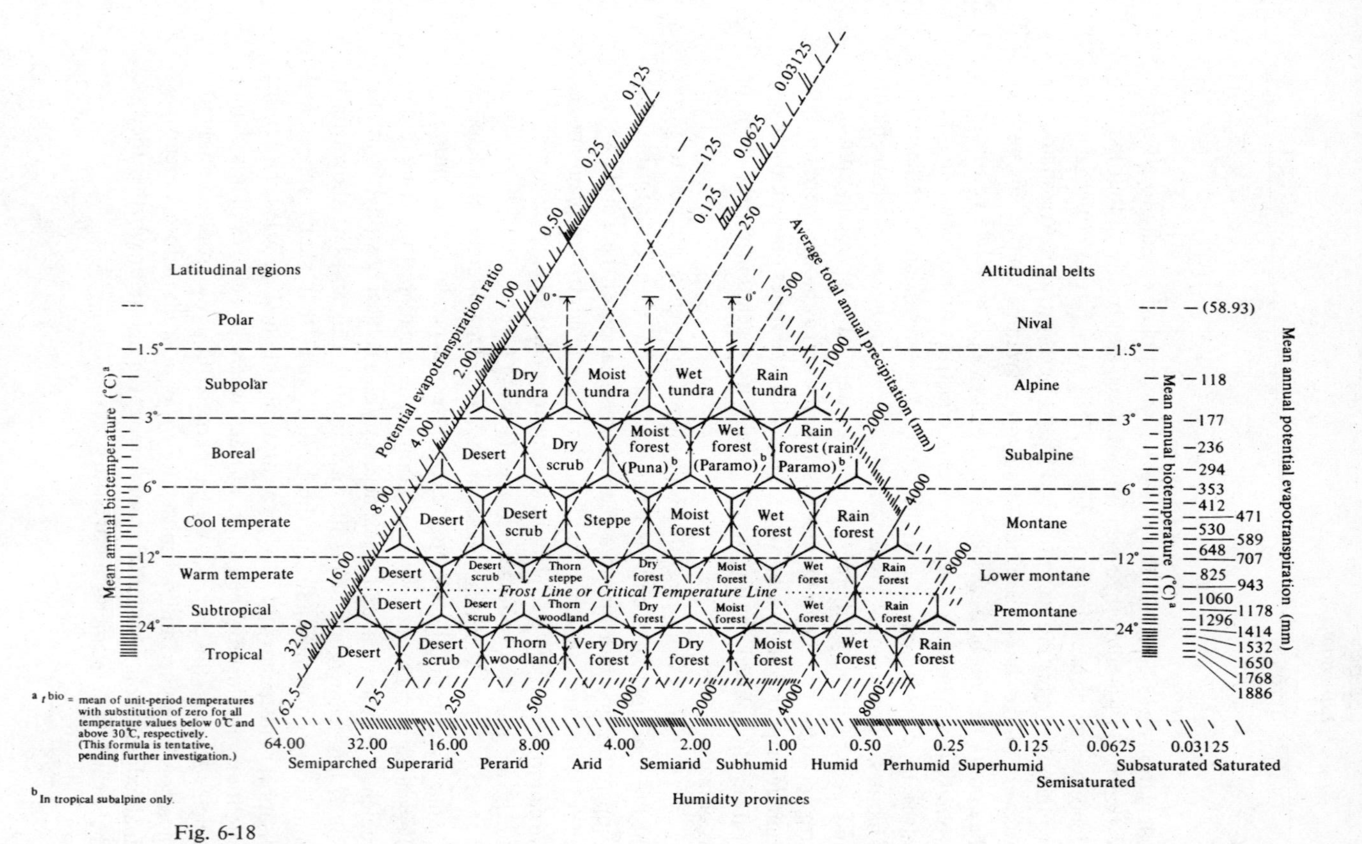

Fig. 6-18

Diagram of the classification of world life zones, or plant formations. (From Holdridge, 1967.)

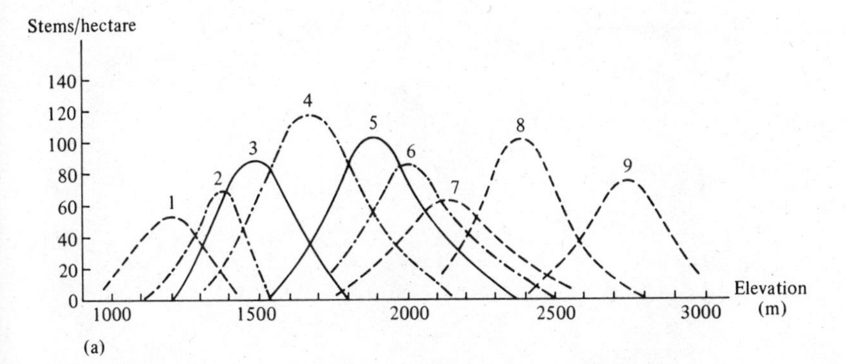

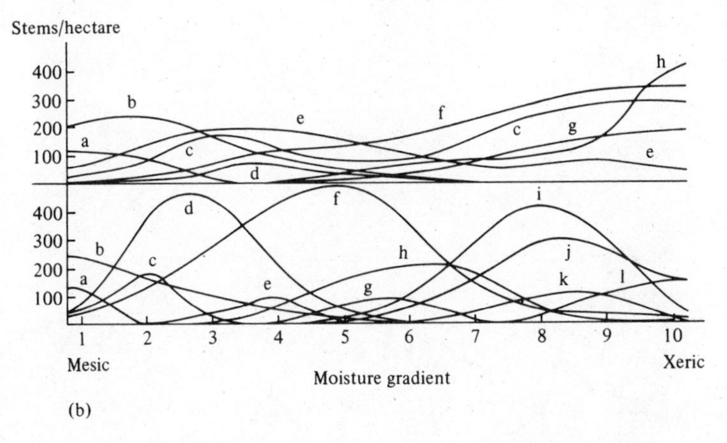

Fig. 6-19

(a) Distribution of oaks and other broad-leaved trees along the elevation gradient in the Santa Catalina Mountains, Arizona. Species shown with solid lines are evergreen oaks of the black oak subgenus *Erythrobalanus*, those with dot-and-dash lines are evergreen oaks of the white oak subgenus *Lepidobalanus*, and those with dashed lines are other broad-leaved species. Species shown: 1, *Vauquelinia californica*; 2, *Quercus oblongifolia*; 3, *Q. emoryi*; 4, *Q. arizonica*; 5, *Q. hypoleucoides*; 6, *Q. rugosa*; 7, *Q. gambelli*; 8, *Acer grandidentatum*; and 9, *A. glabrum*. Although there is a conspicuous "oak zone" on these mountains, the species enter asynchronously. (From Whittaker, 1969.) (b) Population curves for tree species along topographic moisture gradients in mountains. Top: Siskiyou Mountains, Oregon, 460–760 m elevation. Bottom: Santa Catalina Mountains, Arizona, 1830–2140 m elevation. Each letter is a different tree species. Notice that the Oregon species are found over a wider range of conditions. (From Whittaker, 1969.)

Appendix
Delayed Germination of Desert Annuals

As a topic which is important chiefly because of the evolutionary issues it raises, we go more deeply into Dan Cohen's theory of delayed germination of desert plants. Suppose there are N seeds this winter of which a fraction G germinate with the first rain. In good years more rains follow and each of these NG germinating seeds produces S seeds; in bad years, however, there isn't enough subsequent rain and all the NG germinating seeds fail to produce. In both kinds of year the $(1 - G)N$ seeds that fail to germinate remain for the following year. Cohen modifies this to allow some to decay if they stay in the ground an extra year, but this added realism complicates the arithmetic and obscures the evolutionary issues. Suppose a fraction p of the years are good and the remaining $1 - p$ of the years are bad. To summarize, if this is a good year, there will be NGS new seeds left plus $N(1 - G)$ that failed to germinate, while in a bad year there will be only the $N(1 - G)$. In each good year the number, N, of seeds is thus multiplied by $GS + 1 - G$, and in each bad year it is multiplied by $1 - G$. In a long sequence of T years Cohen assumes there will be just pT good ones and $(1 - p)T$ bad ones, so that after T years the N seeds will be multiplied by $(GS + 1 - G)^{pT}(1 - G)^{(1-p)T}$. The average increase per year will be the Tth root $(GS + 1 - G)^p(1 - G)^{(1-p)}$, and we wish to find the value of G, the fraction germinating, that will maximize this rate of increase of the seed supply. Equating the derivative to zero to find the maximum, we find $0 = (S - 1)p(GS + 1 - G)^{p-1}(1 - G)^{1-p} - (1 - p)(GS + 1 - G)^p (1 - G)^{-p}$, and cancelling and solving for G we get $G = \dfrac{Sp - 1}{S - 1}$, which is usually near to p. Suppose $p = \frac{1}{4}$ and $S = 100$ seeds. Then $\dfrac{Sp - 1}{S - 1} = \dfrac{24}{99}$, which is near to $p = \frac{1}{4}$. In this case only about $\frac{1}{4}$ of the seeds should germinate. If fewer do, then the plant isn't reproducing its fastest; if more do, the risks of large scale mortality are too great.

This is Cohen's solution and is doubtless near the truth, but notice its problems: Cohen assumes that of our large number T of years, just pT are good and $(1 - p)T$ are bad. This is like assuming that in a large number of coin tosses exactly $\frac{1}{2}$ will be heads and exactly $\frac{1}{2}$ will be tails. In fact, while that is the most likely arrangement, any other fractions of good and bad years are possible. We can even calculate the probability, p^T, that all T years are good. This will be a very small number but still positive. Now if all these possibilities are taken into account, a very different optimal value of G emerges. The number of seeds after T years is the product of the growth multiples (either $(GS + 1 - G)$ or $(1 - G)$ depending on which kind of year) of the T years. But the expectation of a product is the product of the separate expectations (by the rules of probability, provided the chances of good years are independent in different years) and the expected growth factor for 1 year is $p(GS + 1 - G) + (1 - p)(1 - G)$, so we could take the Tth power for T years. Adding terms, the expected growth per year is $pGS + p(1 - G) + (1 - p)(1 - G) = pGS + (1 - G) = G(pS - 1) + 1$ since $p + (1 - p) = 1$. If $pS > 1$ this increases as G increases and thus reaches its maximum when $G = 1$ (irrespective of p provided S is large enough so that $pS > 1$). Thus, if good and bad years occurred in just the most probable proportions, the plant would be expected to have $G = \dfrac{Sp - 1}{S - 1}$ of its seeds germinate, while if the good and bad years occurred in all possible proportions, the expected populations of seeds would grow fastest with $G = 1$, all seeds germinating. What will the plant really do?

Perhaps the easiest answer is to picture thousands of separate deserts, each like the others except that its climate is determined independently, so that p^T of the deserts have T successive good years, $Tp^{T-1}(1 - p)$ of the deserts have just one bad year, and so on to the fraction $(1 - p)^T$ of the deserts, which have all years bad. If we add the seeds from all the deserts and maximize that sum, then we in fact want the solution $G = 1$. Why is this best? Because the prodigious seed production from that very small fraction of deserts with all T years good outweighs the extinction of our plant from all of the other deserts. What if the fraction p^T is so small that there aren't enough deserts for one, by chance, to have all good years? Then this "best" decision will have been catastrophically bad. If we look at a typical surviving plant in a typical desert, what strategy will we expect of it? Certainly not the "best" strategy we just discussed, for all of those plants would have been exterminated by the bad years. In fact, if we view each desert as an independent case of evolution, each will have its appropriate strategy. Those excessively rare deserts with all good years will favor $G = 1$ (although next year may be disastrous). Those deserts with observed fraction Y of good years will favor the Cohen strategy $\frac{YS - 1}{S - 1}$, and since a large fraction of deserts will have about p good years, a large fraction of the plants we observe will have been best off if they only allowed $\frac{pS - 1}{S - 1}$ of their seeds to germinate.

Let us summarize the results: Picture many random mutations for different G values. Those for $G = 1$ will persist only in rare patches with long accidental sequences of good years, but they will be very common there; those with smaller G will be more and more widespread, and those with $G = \frac{pS - 1}{S - 1}$ will be those we expect to find in any preselected place. Hence, no single solution $G = 1$ or $G = \frac{pS - 1}{S - 1}$ is uniformly best, but Cohen's solution $G = \frac{pS - 1}{S - 1}$ is certainly the most often useful. We can make the new theoretical prediction that the larger the value of G, the spottier the distribution of the desert annual and the denser it will be where it is found. This has never been tested.

Still one more comparison of the strategies can be made. We saw that a plant with $G = 1$ would become commoner and commoner, but more and more local, while those with $G = \frac{pS - 1}{S - 1}$ would persist nearly everywhere. Hence, in a randomly selected patch of desert we were nearly sure of finding plants that had adopted the strategy $G = \frac{pS - 1}{S - 1}$. But now suppose the $G = 1$ plants were very good dispersers and made their way over the other deserts every year. Then again these $G = 1$ plants would be common and widespread. So we have another prediction: Plants with very low dispersal powers will tend to have $G = \frac{pS - 1}{S - 1}$; plants with very high dispersal powers will have larger G, even tending toward $G = 1$.

Patterns of Species Diversity

7

There are more species of intertidal invertebrates on the coast of Washington than on the coast of New England, more species of birds breeding, and also more wintering, in forests than in fields, more species of diatoms in unpolluted streams than in polluted ones, more species of trees in eastern North America than in Europe, and more flies of the family Drosophilidae on Hawaii than anywhere else. There is an even more dramatic difference in the number of species in the tropics than in the temperate (the discussion of which is reserved for the next chapter). Will the explanation of these facts degenerate into a tedious set of case histories, or is there some common pattern running through them all?

A very brief review of explanations that naturalists have proposed will help put the discussion in perspective. One explanation that has been offered is (a) that there are more species where there are more opportunities for speciation and that the presence of many species simply reflects the head start that some areas have over others. Another is (b) that many species occur where fewer hazards have occurred, and that areas with few species have lost species through catastrophes of history. Others explain that (c) there are more species where competitors can safely be packed closely and that the numbers of both species and competitors will not increase with time. Climate, too, has been used to explain diversity; some claim that there are more species where (d) the climate is benign, others where (e) the climate is more stable. It has been suggested that (f) there are more species where the environment is complex and therefore more readily subdivided. Others suggest (g) that there are more species where the environment is more productive. Finally, the abundance of predators has been cited because (h) heavy predation puts a low ceiling on the abundances of separate species, thus allowing more species to fit in, and (i) predators sweep an area clean, leaving it ripe for recolonization by different species.

Some of these explanations are almost meaninglessly vague (what is a benign environment?), and nearly every explanation has been proposed with some particular organisms in mind; but most are plausible and could under some conditions alter the numbers of species. How can we make order out of such chaos? Here is where theory is useful; it can exhibit the roles played by each factor in combination with the others.

Theory of Species Diversity

We will give a very elementary theory relating diversity to the various ingredients by which it can be measured, and we will use the theory to reconcile the various proposals. In its most elementary form we use three ingredients. (1) We need a coordinate representing the range of resource subdivided among the species. For example, we picture height above the ground as this coordinate; we suppose food is available in the foliage from ground level up to the top canopy at 100 ft. Hence we would have a resource line 100 units long. If the canopy were only 40 ft above the ground, the resource line would be only 40 units long. In general, we call it R units long. In another situation foragers might subdivide food by size. In this case food size would be the resource coordinate. If the smallest food were 1 mm long and the largest were 75 mm long, then foods would vary from 1 to 75 mm in length, so $R = 75 - 1 = 74$ units. (2) We need a shorter line for the part of this resource coordinate used by a given species. If a bird feeding in the forest with the 100-ft-high canopy uses only the portion between 5 and 25 ft, we can represent that utilization by a bar (of length $U = 25 - 5 = 20$) between 5 and 25 on the resource axis. (3) We assume there are enough species in combination to utilize every height of resource and that adjacent ones overlap as represented in Fig. 7-1. We split the overlaps, as shown in the figure, so that the segments between the dashed vertical lines sum to the total resource R. The number of species N is, of course, the number of segments between vertical lines so $R = \sum_{i=1}^{N} H_i = N \cdot \frac{1}{N} \sum_{i=1}^{N} H_i = N\bar{H}$ where $\bar{H}$ is the mean distance between vertical lines. Therefore, $N = \frac{R}{\bar{H}} = \frac{R}{\bar{U}} \cdot \frac{\bar{U}}{\bar{H}}$. Since the U's are composed of an H part and O parts, we can write $\sum_{i=1}^{N} U_i = \sum_{i=1}^{N} H_i + 2\sum_{i=1}^{N-1} O_{i,\,i+1}$, and dividing by N,

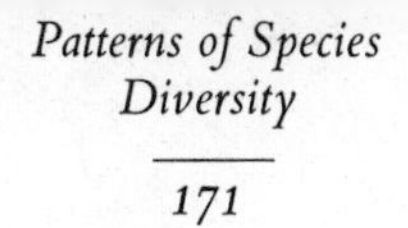

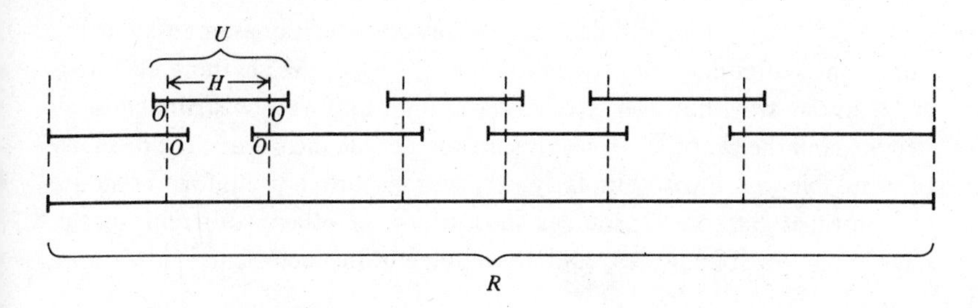

Fig. 7-1

The length R of a spectrum of resources, the utilization, U, per species, and the utilization overlaps. The overlaps are divided into two O's of equal length. See the text for definitions of H and O and for a discussion.

$\overline{U} = \overline{H} + 2\overline{O}$. (Notice, however, that $\overline{O}$ is not the ordinary average of the lengths of the overlap segments; it is the average, per species, of these lengths, and the end species only overlap with one species while the middle species overlap with 2. Hence, our $\overline{O}$ is slightly less than the strict average of the separate O's.) Substituting this into $N = \frac{R}{\overline{U}}\frac{\overline{U}}{\overline{H}}$ we find $N = \frac{R}{\overline{U}} \cdot \frac{\overline{H} + 2\overline{O}}{\overline{H}} = \frac{R}{\overline{U}}\left(1 + 2\frac{\overline{O}}{\overline{H}}\right)$. This is just arithmetic and contains no biology; it simply relates the lengths of the overlapping U segments to the line R they fill and the necessary overlaps. It describes the obvious fact that you cannot increase the number N of U segments (i.e., of species) without increasing the relative overlap $\frac{\overline{O}}{\overline{H}}$ or decreasing $\overline{U}$ or increasing R. This same picture applies if the resource is subdivided in more than one dimension. If, for instance, both resource height and size are subdivided, R is a rectangle with length the range of resource heights, and width the range of resource sizes. Each species utilization is also a rectangle and we have enough overlapping rectangles to cover R. Now there are more neighboring species, 4 or more, depending upon the arrangement of rectangles, so we expect

$$N = \frac{R}{\overline{U}}\left(1 + C\frac{\overline{O}}{\overline{H}}\right) \tag{1}$$

where C is some number measuring numbers of neighbors.

This formula is necessarily correct; it does not assume the importance of competition or predation or history or anything else. It is crude in one way, however. Resources are not uniformly distributed at all heights, and the range, R, of heights is thus a crude measure of the diversity of available resources. Similarly, utilizations are not uniform but are concentrated on some resources more than on others, different species have different abundances, and the competition coefficient α is a more suitable measure of overlap than $\frac{\bar{O}}{\bar{\bar{H}}}$. Alpha is defined precisely in the Chapter 2 appendix; roughly, it is the harm done by an individual to a competing species compared to the harm done by a member of the competing species to its own population. It is pleasant, and a little remarkable, to find that a strictly analogous formula holds when we substitute more practical measures of diversity of the production of resources and of utilization. As before (p. 113), we use the formula $\frac{1}{\sum_i p_i^2}$ as a measure of diversity. If the p_i are the proportions of the total resources in height interval i in the tree, then $\frac{1}{\sum_i p_i^2}$ is called the diversity of resources, D_R; if p_i is the proportion of a species' resource utilization in this height interval, i, then $\frac{1}{\sum_i p_i^2}$ is the diversity of utilization, D_U; if p_i is the proportion, of all of the individuals, that belongs to the ith species, then $\frac{1}{\sum_i p_i^2}$ is the species diversity, D_S. Then it is true that

$$D_S = \frac{D_R}{D_U} \cdot \lambda \tag{2}$$

and λ (called the Rayleigh ratio of the α matrix) is not far from $1 + C\bar{\alpha}$ where $\bar{\alpha}$ is the mean competition coefficient and C again measures the number of neighbors and depends upon the number of subdivided dimensions. This is technically somewhat better than Eq. (1), but its intellectual content is the same; hence, we postpone its lengthy proof to the first appendix to the chapter.

What is the subdivided resource continuum? Is it really height above the ground or food size, or both, or neither? Only the naturalist familiar with a group of organisms is prepared to guess R or D_R. He can often test his guess by the tidiness of the results, as we shall see. Equation (1) or (2) is correct even if he uses the wrong measure of the resource continuum, for as we said before, they contain no biology. But the biologist who guesses the right resource coordinates for his species gets a special bonus: the R, or D_R, is then independent of the overlap $\frac{\overline{O}}{\overline{\overline{H}}}$, or α; the overlap may then measure the "true" overlap experienced by the species. Suppose, for example, that birds are really subdividing food by both size and height in the tree, but that the biologist is guessing that only height in the tree is relevant. Accordingly, he observes a large amount of feeding height overlap, an amount which is actually greater than exists since, although many species are feeding at the same height, they are eating foods of different sizes and thus reducing their overlap. So far so good. Now suppose the biologist compares a second habitat of about the same foliage height but with a smaller range of food sizes. He then records the same R but finds fewer species and less overlap. Here he blames a change in the species overlap for the reduction in the numbers of species, while if he had guessed the entire appropriate food spectrum, he would have observed the same overlap but would have found reduced R in the second habitat. This would be a useful finding because then the overlap could be viewed as a biological property of the species and R as a measure of the environment; but these interpretations of overlap and R hold only if the biologist has guessed correctly.

In what follows, we use Eqs. (1) and (2) to interpret the various species diversity patterns.

History or Equilibrium: The Principle of "Equal Opportunity"

Some problems in numbers of species truly seem to be outcomes of a capricious history. That is, they are interpretable in historical terms and not in terms of the machinery controlling species diversity.

In short, our Eqs. (1) and (2), although applicable, are irrelevant in cases where history is paramount.

For example, the fossil records suggest that Europe and North America had roughly equal numbers of tree species prior to the Pleistocene glaciations. During the glaciations Europe lost a large fraction of tree species while North America lost virtually none, and the time since then has been insufficient for Europe to reacquire the species it formerly had. Hence, Europe is impoverished in trees for historical reasons. Presumably, the actual machinery involved was that the advancing glaciers forced the trees to move south ahead of them. In eastern North America where the mountains are in a north-south chain and form no barrier, the retreating tree species were always able to continue south to areas where climates were suitable. In Europe where the Alps and the Mediterranean lie east-west and therefore form barriers to southward movement, warm-loving tree species being pushed south by glaciers were squeezed in a vanishing warm zone between cold glaciers and cold mountains or cold glaciers and the sea. These trees found no refuge and went extinct. History, pure and simple, appears to be the only explanation for the difference between European and American tree diversities.

However, if we study the birds breeding in these very forests we get different results. If we compare a few acres of forest in Europe with similar acreage in North America, we find about 20 species in both habitats; and we get equivalent results if we contrast other habitats in Europe with similar habitats in North America. A small field in both places would have only a handful of bird species breeding, but a bushy field would have more, and a forest of many layers would have even more. Hence, in the comparison of similar habitats in North America and Europe, tree species diversity is explained by history; bird species diversity, which does not differ between habitats of similar structure but between habitats of differing structure, is explained by those habitat differences. Of course, birds are much more mobile than trees, and history has apparently left little trace on the comparison of numbers of bird species between Europe and North America.

Thus, for birds at least, we begin to look for explanations in terms of how many bird species a habitat will "hold." A forest, for some reason that we wish to investigate, will "hold" more bird species than a field. This is not to imply that both are "full" and cannot possibly receive another. Rather, it is more like a gas that equalizes its pressure between two connected vessels, the larger of which automatically equilibrates with more gas. This analogy can be made more precise with what may be called the principle of "equal opportunity." We picture two habitats, A and B, connected by a corridor permitting free migration back and forth (Fig. 7-2). We also assume that the species can rapidly and easily adapt to either habitat upon migration, but that changes are necessary, and that a single species can seldom simultaneously occupy both. Finally, suppose there is a large enough total number of species so that when they are all together, they compete rather severely. Then the benefit of an A → B migration might fall off as shown in Fig. 7-2, where p is the proportion of species in habitat B such that both habitats are equally good as far as a potential migrant is concerned. When B has fraction p of the species, both habitats offer equal opportunity for further colonization and the exchange stops. As further species enter the area, both habitats may acquire new species but the balance is always maintained between the two, provided the entry of new species is slow compared to the shifting between habitats.

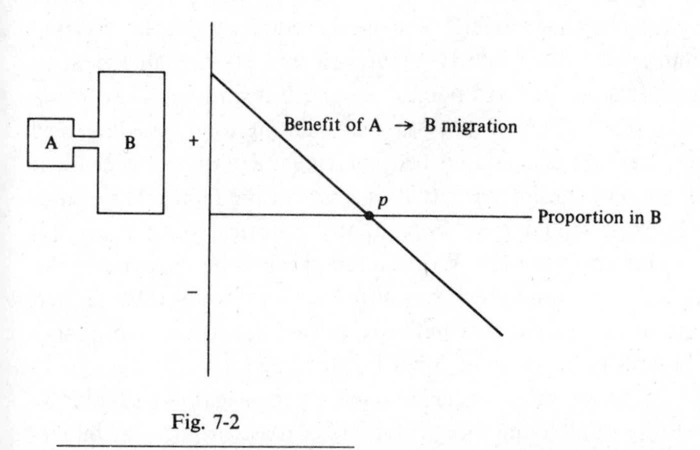

Fig. 7-2

The principle of equal opportunity. A and B are two different habitats of unequal size, joined by a corridor as shown on the left. On the right is portrayed the benefit of A-to-B migration as a function of the proportion of the species in habitat B. When this line drops below zero, B → A migration is favored. Where the line crosses at proportion p, the habitats have equal opportunity for further colonization.

In the case where the only differences between the habitats are in the lengths of the R lines, and possibly in the overlap between the species utilizations, we can be perfectly precise about what equal opportunity means. Then the species K values would be the same in either habitat, so the species will prefer the one with less competition. Equal opportunity will mean equal $\frac{\overline{O}}{\overline{H}}$ or equal α. Thus, two equally productive habitats that might be expected to have equal K's for species will be expected each to contain that number of species that will maintain an equal α in the two habitats. If one habitat temporarily has smaller α, there will be a net flow of species entering this habitat, restoring the equality of the α's. When different resources have different productions, then K's as well as α's will vary and it is not so easy to say just when the migrations will cease. However, there will still be a time, and a proportion p in habitat B, at which further migration will cease or balance because the habitats have equal opportunity. In such cases, nearby habitats should reveal a pattern of species diversity, a pattern describable in terms of Eqs. (1) and (2), or something similar.

There is another kind of historical pattern that we can avoid if we wish. Hawaii has more flies of the family Drosophilidae than anyplace else, perhaps because it has fewer flies of other kinds and fewer other ecologically similar insects. The clever entomologist by plotting the total number of fly-like insects might well find Hawaii equivalent to other tropical areas (or perhaps poorer because it is an island). To make this clearer we recall (from p. 174) that by plotting total breeding bird species, differences between habitats became regular and quite independent of history. If we had considered only flycatchers of the family Muscicapidae, the differences would have been purely historical since those flycatchers are confined to the Old World and appear to be replaced by the tyrant flycatchers of the family Tyrannidae in the New World. If we plotted all flycatching birds, the patterns might reappear; but it would probably be safer to plot all birds, or some other larger category.

At the other extreme people often suggest we should explain the diversity of all living things, not just of trees, or birds, or butterflies. But to suggest this is not only a little masochistic because the counting job would be virtually impossible—it also misses the point. We are looking for general patterns, which we can hope to explain. There are many of these if we confine our attention to birds or butterflies, but no one has ever claimed to find a diversity pattern in which birds plus butterflies made more sense than either one alone.

Hence, we use our naturalist's judgment to pick groups large enough for history to have played a minimal role but small enough so that patterns remain clear.

Habitats Differing Only in Structure

We have seen that if the biologist is clever, he measures R so that it reflects the environment and U so that it is a measure of the species' abilities determined as in Chapter 3 on economics, and overlap $\frac{\bar{O}}{\bar{\bar{H}}}$ becomes a true measure of the species' interactions. It is very plausible then (recall the principle of equal opportunity) that nearby habitats would be occupied by species of the same U and same mean overlap $\frac{\bar{O}}{\bar{\bar{H}}}$, and differ only in their R values. Comparing the numbers of species in such habitats should reveal that the number of species is proportional to the range of resources (Eq. (1)), or, using Eq. (2), that the diversity of species is proportional to the diversity of resources.

Cases of this sort have been explored in birds, first by R. and J. MacArthur (1961). These authors used a slightly different diversity measure, so we here (Fig. 7-3) replot some old plus some new data to show breeding bird species diversity, D_S (calculated in homogeneous census areas large enough to hold about 25 pairs), against "foliage height diversity," which seemed the best estimate of the real D_R. For calculating foliage height diversity, $\frac{1}{\sum_i p_i^2}$, p_1 was the proportion of the total foliage in the layer of herbaceous vegetation, p_2 the proportion of foliage in the layer of bushes and understory, and p_3 the proportion in the canopy. The general linear arrangement of the points suggests that R variations—variations in arrangement of foliage layers—are responsible for most of the variation in bird species diversity. This is not proof that foliage height is the only resource subdivided; there surely are others, such as food size,

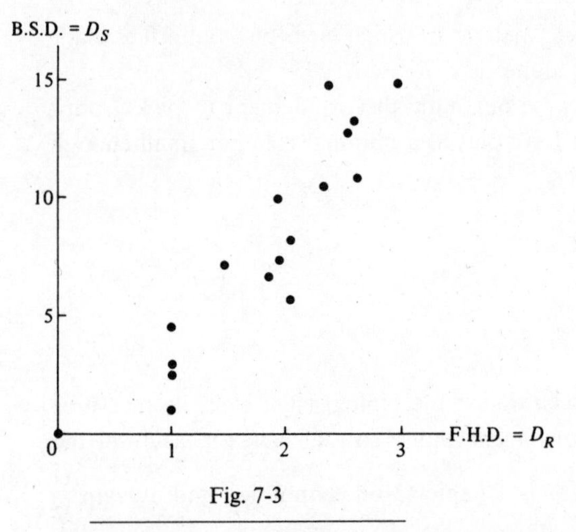

Fig. 7-3

Bird species diversity, B.S.D., is plotted against foliage height diversity, F.H.D. (the number of equally used layers of foliage). F.H.D. is a measure of the length of R.

but apparently nearby habitats differed in foliage height and not in food size spectrum, so the differences in numbers of bird species were controlled by the differences in foliage profile. The MacArthurs also showed that in eastern deciduous forests knowledge of the number of plant species was irrelevant to the prediction of the number of bird species. Thus, a pure red maple forest of a given foliage profile would have as many breeding bird species as a mixed forest with the same profile.

This result does not prove that bird species recognize their habitat by profile alone; many may actually require a particular tree species for food or nest site. Rather it is the *number* of species that depends upon the profile, and if one disappears because some tree species disappears, another replaces it.

In future studies some better measure of the range or diversity of resources will doubtless be devised, perhaps based on direct sampling of the food. Such studies may improve the clarity of the relation between the number of species and the range of resource.

Structure of the habitat may affect more than just R. Cody (1968) showed that in dense wet grassland of medium height, bird species are hard pressed to subdivide food. Different species end up selecting patches of slightly different density or wetness. Thus, Leconte's sparrows were in the wettest fields. But when he studied taller grass fields, the birds had used a new dimension. Not only do some species select denser or wetter areas, but some feed on the ground and others from the grass tops. There is now a vertical as well as a horizontal dimension of their subdivision. In terms of Eq. (1) or (2), the constant C has increased in the tall grass areas, and consequently more species could be present. Although they are less well worked out, there must be innumerable cases where added geometrical structure gives added dimensionality to the species subdivision and hence increases C.

Patrick (1968) compared the diatom floras developing independently in separate boxes filled with water from the same uniform source. Thus she tested for the role of accident in species diversity. In fact she found that in the absence of variation in R or production and under replicated conditions the diatom floras were very similar. Over 95% of the individuals were of species shared in common among the boxes, and other measures of diversity were very similar also. Hence, diversities are similar when conditions are similar.

Patrick (1963) also contrasted the numbers of species of diatoms in unpolluted rivers of different chemistry. Remarkably enough, provided the rivers were of complex structure the numbers of species (but not their names) were remarkably uniform. Relatively unstructured springs and other simple habitats had fewer species however.

Habitats Differing in Climatic Stability

There are two ways in which the fluctuations of climate would affect Eqs. (1) and (2). First, a species adapting to a varying environment must have a large U, whereas in a constant environment U can be smaller. For example, in a wet tropical environment fruit is available throughout the year and many birds can, and do, specialize on it. In more seasonal places, fruits are only sometimes available and a resident species must enlarge its range of utilization, U, until it includes some items for every season. Since U appears in the denominator, increase in climatic fluctuations causing increasing U will decrease the number of species. For the same reasons competing species in a fluctuating environment tolerate less overlap (Chapter 2), so that we may also have reduced $\frac{\bar{O}}{\bar{H}}$ in a fluctuating environment. For this reason, too, we expect fewer species, although this reason applies only to competitors.

The best account of the role of environmental stability is due to Sanders (1968, see also Sanders 1969, Slobodkin and Sanders, 1969). Since his technique is interesting in its own right we discuss it with some care. Most ecologists interested in diversity have wanted a number or a few numbers describing aspects of diversity (e.g., one number for number of species and another for degree of dominance of common species over rare; others have used the statistics of a log-normal distribution fitted to relative abundance data, etc.). They then use these numbers in regression analyses or in graphs to get numerical relations between diversity and measurable aspects of the environment. Sanders, however, describes diversity with a function, not just one or two numbers. In this respect, of course, he includes more information in his measure of diversity, but he does sacrifice the ability to plot diversity numerically on a graph and contents himself with qualitative comparisons.

Basically Sanders plots a curve of the expected number of species in random subsamples of different sizes. If there are, say, 1000 individuals in his total sample, he calculates (by a crude but effective approximation) the expected number of species in each size of subsample and plots these as in Fig. 7-4. Each curve represents a "species-individual" function calculated from a single sample of bivalves and polychaetes from marine benthic sediments. Specifically, suppose there are n_i $(i = 1, \ldots, N)$ individuals in the ith of the N species and that $\sum_{i=1}^{N} n_i = M$, the total number of individuals. Sanders then plots the point (M, N) on the graph. To get the interpolated, "rarefaction" points connecting this to zero, Sanders proceeds as in the following example. To get the value of the curve over, say, 25 individuals, Sanders lets $S = \frac{25}{M}$ be the fraction of the total individuals in the subsample of 25. Any species i for which $Sn_i \geqslant 1$ he assumes is automatically in the subsample. Let $R = \sum n_j$, summed over all those j such that $Sn_j < 1$, be the number of individuals in species that are too rare to be guaranteed present in the subsample. Then Sanders wants a fraction S of these individuals from rare species in the subsample.

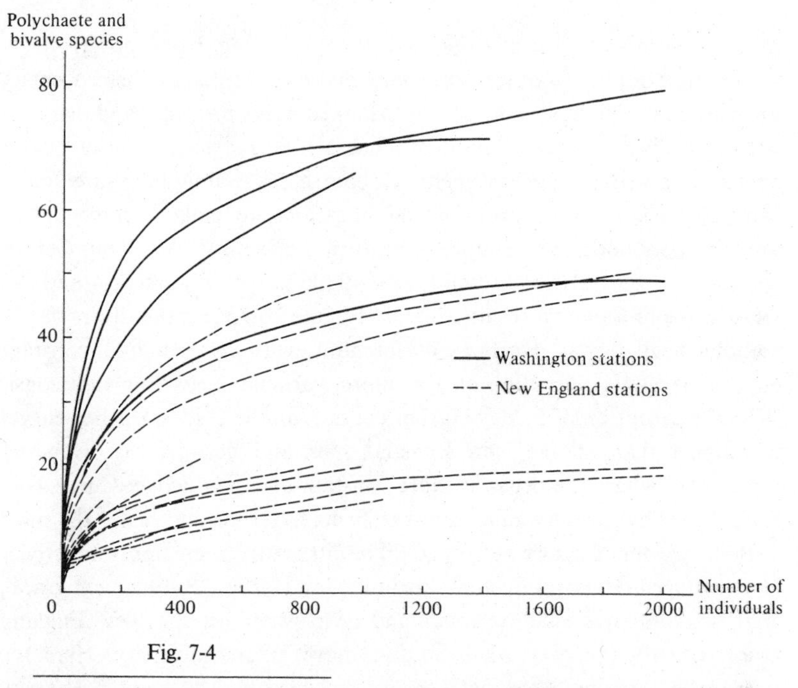

Fig. 7-4

Comparison of diversity values for the Friday Harbor, Washington, and southern New England series of stations. See the text for a discussion. (From Sanders, 1969.)

He assumes they are found one individual per species so he assumes the subsample has *SR* rare species. Hence, in total, the number of species in the subsample is *SR* plus the number of species with $Sn_i \geqslant 1$. This is a rough approximation to the expected number of species in a random sample of 25 individuals.

The trouble with treating the curve itself as diversity is that one can't in general call one curve "greater than" another curve the way one can call one real number greater than another. However, in most cases Sanders' curves don't cross. Then one curve is uniformly above another and it is reasonable to say that the former represents an area of greater diversity.

One final warning: Sanders' curves are not the same as the species-individual curves one would get by counting progressively larger real samples. Instead, they are roughly the species-individual curves one would get by stirring up all of the individuals in the whole area into a random homogeneous mixture and then counting from progressively increasing samples. In other words, any microenvironmental heterogeneity of the mud in Sanders' samples is overlooked in his measure, and one area with more heterogeneity would produce a higher diversity curve than a similar area with less heterogeneity. Of course, his technique of rarefaction can be applied, with appropriate modifications, to make a curve out of any diversity measure: Simply substitute "diversity" for "number of species" on the vertical ordinate axis of the graph. Figures 7-4 and 7-5 show some of Sanders' results. The first figure compares the diversities of stations near Friday Harbor, Washington, with its rich and constant conditions, with diversities on the more variable New England coast. With the exception of Friday Harbor curve 1, all the Friday Harbor curves are higher than all the New England ones, and we infer the diversities are greater where the ocean is more constant and productive. Curve 1, in fact, is the shallowest water station at Friday Harbor and hence the most variable. So far, Sanders' results could be either due to greater productivity or to reduced temporal fluctuations in Friday Harbor. But his next graph, Fig. 7-5, compares shallow water and deep water off the New England coast. Clearly the deep-water stations have greater diversity. Here the deep waters are less seasonal and less productive. The greater diversity can then only be due to the reduced seasonality or other temporal fluctuations in deep water. Sanders reports that the shallow-water stations have up to 23°C seasonal temperature change whereas waters as deep as 300 meters have only a 5°C seasonal change. Thus the reduced seasonality is the one factor he always finds associated with greater diversity. It seems inescapable that the reduced seasonality has caused an increased diversity. But in the absence of more numerical data we cannot rule out a subsidiary effect of productivity. To rule out productivity by his methods, Sanders would need to compare areas with identical seasonalities but very different productivities.

Sanders does not attribute all of diversity to climatic stability. He also says areas that subject their organisms to reduced "stress" have greater diversity. Unfortunately, though seasonality has some measurable objectivity, stress is harder to specify. Is reduced productivity a form of stress? If so, Sanders implies productivity affects diversity, but he seems to deny this. In any case his data are of extraordinary interest and we return to them briefly in Chapter 8.

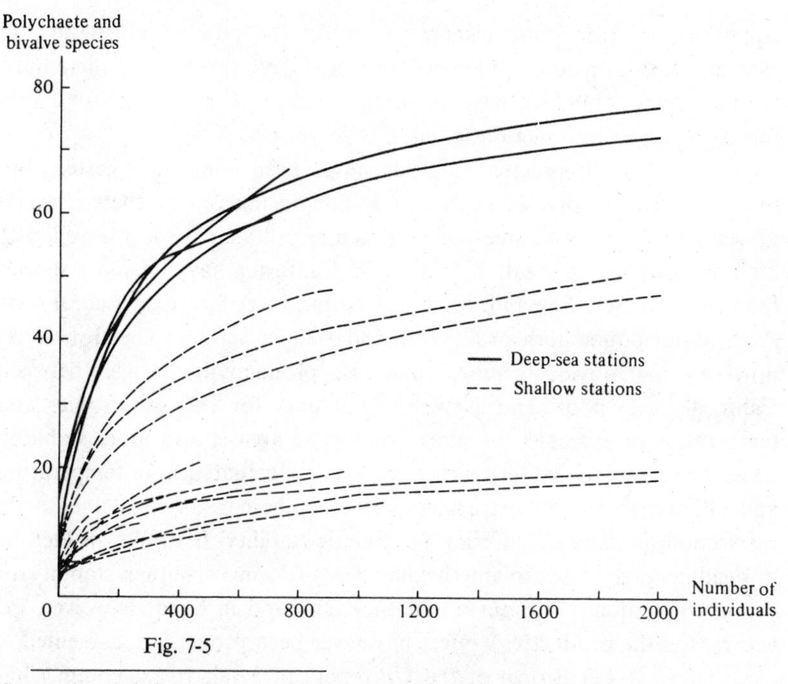

Fig. 7-5

Comparison of diversity values for the continental shelf and slope stations of the southern New England transect. (From Sanders, 1969.)

Habitats Differing in Productivity

The knowledge that two habitats differ in their total productivity is not sufficient information to allow us to predict whether one will have more or fewer species. If every resource has uniformly increased production in one habitat, that habitat should have greater R and reduced U; the R will be greater because resources that were formerly too scarce to form an adequate diet and therefore not counted as part of R now allow a species to survive and are counted. U will be less because, as we proved in Chapter 3, species should have a more specialized diet where food is denser. Since food is often denser where production is greater, we conclude U will often be reduced. Both growth in R and reduction in U will cause an increased number of species, according to the equations, so production clearly can affect the number of species. In extreme cases it obviously does: where there is zero production, there must be no species. Thus, we have theoretical reason to expect areas of uniformly increased production to support more species.

In practice this has never been adequately tested, but there is a general correlation that looks convincing. Where there is ocean upwelling on the west sides of continents, productivity is marvellously high and diversity is great. Tide pools in California have far more species than those in New England. Similarly, tropical wet forests and coral reefs are without doubt both productive and rich in species. The trouble is, however, that obviously more than just productivity changes between California tide pools and New England ones, or between tropics and temperate. For example, the more productive area is also more stable in time. The water in California tide pools may fluctuate less in temperature and salinity, so that its extra species can also be partially explained as in the section on habitats differing in climatic stability. It would, in fact, be misleading to try to explain the increase in terms of either stability or productivity alone. Without doubt they act hand in hand. However, for this reason the productivity effect has never been properly documented.

J. Brown of the University of Utah (pers. comm.) has discovered one case in which productivity may control species diversity. He finds that the species diversity of the heteromyid rodents (kangaroo rats and pocket mice) that occupy patches of desert in Nevada is greater where the rainfall is greater, even though there is no conspicuous difference in vegetation in these rainier patches. Brown conjectures that the heavier rainfall increases the seed production of the plants. The rodents divide resources partly by seed size and partly also by foraging position; some feed under bushes and others feed in the open. Hence it is plausible that the increased seed production would allow enough seeds to fall in the open to make that way of life feasible, whereas the open-feeding rodents would be unable to make a living where seed production is low.

On the other hand, there are many cases where an increase in total productivity actually reduces the number of species. Patrick, Hohn, and Wallace (1954) have compared the numbers and diversity of diatom species on the same river just above and just below where organic pollution was pouring in. The polluted waters were more productive in the sense that they supported more diatom individuals and more rapid photosynthesis, but there were fewer species and lower species diversity. In the polluted waters there were few, but exceedingly common, species; in unpolluted waters, many, but less common, species.

There are several ways of viewing such results. They make use of the fact that the complete spectrum of nutrients is not being enriched by the pollutants, so that some nutrients become, relative to others, very abundant. Diatoms consuming mainly these nutrient resources will become very abundant and will do proportionately more damage to their competing species, which consume mainly the unenriched nutrients. Thus the species will only coexist in the polluted environment if their overlaps, $\frac{\bar{O}}{\bar{\bar{H}}}$, are relatively small. A greater overlap would have been tolerated in the unpolluted stream. Hence increased production, if concentrated on some resources, can reduce $\frac{\bar{O}}{\bar{\bar{H}}}$ and thereby decrease species diversity.

Another, supplementary explanation for reduced species diversity in polluted areas is that if R is not reduced, at least D_R is, for D_R builds in the relative abundance of the resources and is interpreted as the number of equally dense resources. When one nutrient becomes excessively common, it reduces D_R, and for this reason reduces species diversity.

Habitats Differing in Structure, Stability, and Productivity

Most pairs of randomly selected habitats will differ in more than one respect; often they will differ in structure, environmental stability, and in productivity. Then all of the ingredients of Eqs. (1) and (2) may be altered, conceivably in ways that would cause opposing effects upon the numbers of species. In this case, to understand changes in the numbers of species there is no alternative but to measure R, U, C, and $\frac{\bar{O}}{\bar{\bar{H}}}$ and combine them according to the equations. Since people have seldom even measured single components in these equations, it is no wonder no progress has been made in comparing habitats that differ in many ways simultaneously.

Matters of Environmental Scale

A real environment has a hierarchical structure. That is to say, it is like a checkerboard of habitats, each square of which has, on closer examination, its own checkerboard structure of component subhabitats. And even the tiny squares of these component checkerboards are revealed as themselves checkerboards, and so on. All environments have this kind of complexity, but not all have equal amounts of it. In an orchard or creosote bush (*Larrea divaricata*) desert there is not much large scale variation; the big squares of our checkerboard are all identical, if their sides are, say, about 100 meters long. In contrast, 100-meter-squares laid on a field being recolonized by trees would be very different from each other. Some would contain almost solid patches of forest, others would still be grassland, and yet others would be filled with bushes and vines. Our problem is this: Over what size square should we count the number of species? Are the patterns of species diversity more orderly and more accessible if we choose small squares, medium squares, or very large ones? Three fairly obvious points help us to answer.

First, if our squares are too small, they will hold a very poor sample of individuals and species. A square in which we sample only three individuals cannot have more than three species. We must adjust the census technique or choose a square large enough so that the sample is big enough to reveal many species if many are present.

Second, if we choose very large squares, of the size, say, of 2000 miles or kilometers on a side, then we are involved in areas big enough for speciation. Two such squares would include most of eastern United States and most of western United States. The eastern square would have relatively little topographic diversity and hence little opportunity for the geographic isolation used in speciation. Even the Appalachian Mountains are essentially a continuous range, so that there are very few superspecies with different parts of the range containing semispecies. The west is very different. There are innumerable mountain barriers and mountain islands and hence there are innumerable superspecies and species groups. Figure 7-6 shows the numbers of North American chipmunk species of the genera *Tamias* and *Eutamias*, compiled from maps in Hall and Kelson (1959). Even in squares of the size shown on the map, topographic diversity has an evident effect. The east has 1 or 2 species and the west has 14, but these western species are almost always in distinct habitats. There can be no doubt that history, the history of species formation, is revealed even in squares no bigger than those on the map.

Third, if competition is the cause of most patterns of species diversity, then the square that best revealed these patterns would be one small enough so that its species would be coexisting. Hence, we hope for patterns in relatively homogeneous habitats of a size just large enough to hold an adequate sample of species. In squares of this size we have reason to hope the traces of history will have been erased. This is why homogeneous habitats, even in the west, usually or almost always have at most a single chipmunk species.

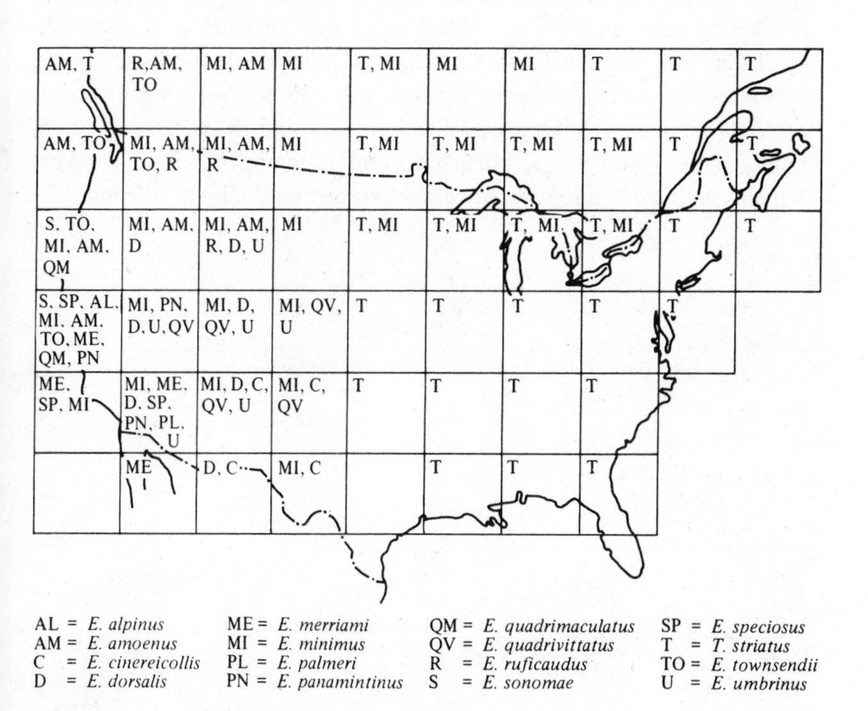

Fig. 7-6

Ranges of chipmunk species showing that areas of the size of the squares have many more species in western United States than in eastern United States.

If we have large squares within which there are many species for historical reasons but small component squares with just the small number of species that can coexist, then we are sure that the component squares must have different species. In this case, the number of species grows rapidly with area sampled, from the number of coexisting species to the total in the large area. An area like the east with very little geographic diversity will have its number of species grow less fast with area sampled (Fig. 7-7).

It is hard to reverse this argument and infer from the growth in number of species with arëa how finely the species have subdivided their habitat. This is difficult because the topographic diversity is confounded with the species subdivision. The species-area curve will be steep if either the area is topographically diverse or the species have subdivided by very subtle criteria. Thus, if we wish to know how subtle the species habitat subdivision is, we must use other methods. Basically, we should compare the difference between the censuses with the difference between habitats. The steepness of this curve would reveal the subtlety of habitat subdivision alone, independent of topographic diversity. What remains is to work out how to measure species and habitat difference.

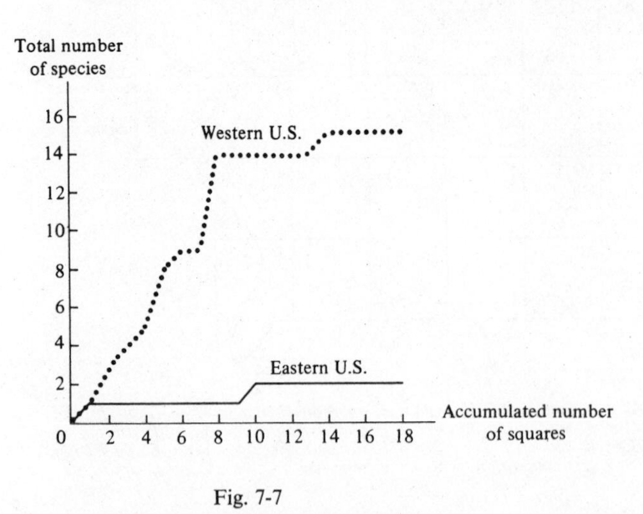

Fig. 7-7

Species-area curves for the chipmunks of Fig. 7-6 The curves were constructed by beginning at the bottom left in the map and working upward, recording new species as they appeared.

There are many suitable measures of the difference between the species of two habitats and of the difference in structure (Horn, 1966). If P and Q are two habitats being compared, and if p_i and q_i are the fractional abundances of the ith species in habitats P and Q, respectively, then $\frac{p_i + q_i}{2}$ is the fraction composed of the ith species in a mixture containing equal-sized samples of both habitats, P and Q. By our former reasoning (p. 113) $\frac{1}{\sum_i \left(\frac{p_i + q_i}{2}\right)^2} = \frac{4}{\sum (p_i + q_i)^2}$ is the number of equally common species in the combined census of both habitats together. If we divide this by $\frac{1}{\sum_i p_i^2}$, the number of equally common species in habitat P, we get the multiple of the number of species in habitat P that P plus Q contain. To be impartial, we also divide it by $\frac{1}{\sum_i q_i^2}$ and average the resulting multiples. The result is the average multiple, M, of the number of species in separate habitats that the combined census contains:

$$M = \frac{1}{2}\left[\frac{\frac{1}{\sum\left(\frac{p_i + q_i}{2}\right)^2}}{\frac{1}{\sum p_i^2}} + \frac{\frac{1}{\sum\left(\frac{p_i + q_i}{2}\right)^2}}{\frac{1}{\sum q_i^2}}\right] = \frac{1}{2}\left[\frac{4\sum p_i^2 + 4\sum q_i^2}{\sum p_i^2 + 2\sum p_i q_i + \sum q_i^2}\right]$$

$$= \frac{2}{1 + \frac{2\sum p_i q_i}{\sum p_i^2 + \sum q_i^2}} = \frac{2}{1 + ov}$$

where ov is a well-known measure of the overlap between the communities. This may seem complicated, but it is really simple. If P has 4 equally common species and Q has 4 more, all different, then there are 8 species in all and $M = \frac{8}{4} = 2$. This is the largest value M can take; the combined census is only double the separate ones if there are no species in common. Suppose P has 4 and Q has 4, but now 2 are shared; then $M = \frac{6}{4} = 1.5$, indicating that the combined census has 1.5 times as many species as the separate ones. Finally, if the 4 species in P and Q are all shared, then M takes on its minimum value $\frac{4}{4} = 1$. The elaborate formula for M is only different in that it takes into account the possibility that not all species are equally common. Suppose, for example, that P has three species of relative abundance $p_1 = 0.25$, $p_2 = 0.25$, $p_3 = 0.50$, and $p_4 = 0$ and that Q has two species of abundances $q_1 = 0$, $q_2 = 0$, $q_3 = 0.50$, and $q_4 = 0.50$; only species 3 is shared. Then $\sum p_i^2 = (0.25)^2 + (0.25)^2 + (0.50)^2 = 0.374$; $\sum q_i^2 = (0.5)^2 + (0.5)^2 = 0.5$; and $2 \sum p_i q_i = 2p_3 q_3 = 0.5$; so overlap, *ov*, equals $\dfrac{0.5}{0.874} = 0.57$. Hence, $M = \dfrac{2}{1.57} = 1.27$. In other words, the combined census has on the average 1.27 times as many "equally common species" as does P or Q.

MacArthur and Recher (MacArthur, 1965) compared pairs of bird censuses in Puerto Rico and also pairs of censuses in the United States. Each pair produces a point in Fig. 7-8, in which differences in bird census and differences in layering of the habitats are compared. (What they call *BS diff* (bird species difference) is essentially the logarithm to the base *e* of our M for differences in bird species, and what they call *FH diff* (foliage height difference) is essentially the $\log_e$ of our M value for layer densities in which p_1 = proportion of herbaceous foliage, p_2 is proportion of bushes, and p_3 is proportion of canopy over 25 ft. $\log_e 2 = 0.693$,

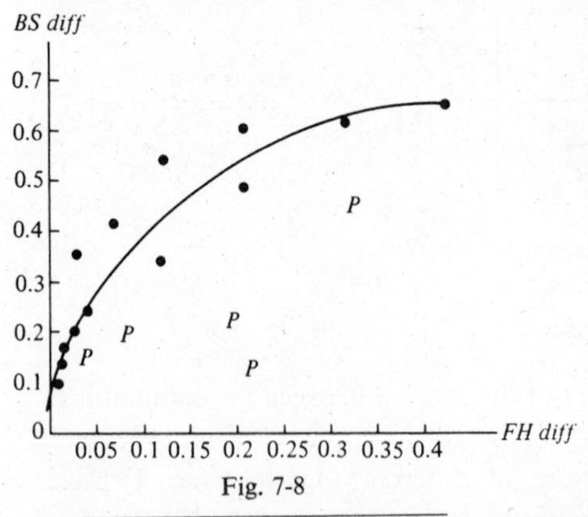

Fig. 7-8

Difference in bird species composition (*BS diff*) between census areas differing in foliage profile is plotted against foliage profile difference (*FH diff*), showing that in Puerto Rico (*P* points) the same difference in habitat causes much less change in bird species than on the U.S. mainland (● points). (From MacArthur, 1965.)

which is the maximum value on the scales of the figure and $\log_e 1 = 0$, which is the minimum.) Clearly the same difference in profiles causes a larger difference in bird species on mainland United States than in Puerto Rico. In other words, different habitat squares in Puerto Rico are more likely to contain the same species than those in the United States. M has been called the "between habitat diversity," and the preceding statement can be rephrased by saying, Puerto Rico has smaller bird "between habitat diversity" than the United States. We already know island species have expanded habitats. This is an exact documentation of the fact.

MacArthur, Recher, and Cody (1966) attempted to extend these results to the tropics with a new set of points of Panama comparisons. But these results were less satisfactory. Their censuses and their measures of habitat appear to have been inadequate.

Pianka (1969), comparing the rich lizard fauna of the Australian deserts to the poorer one of American deserts, concluded that Australian lizards have subdivided habitats more finely. One would infer that a lizard species-area curve would grow faster in Australia. But on top of this greater between-habitat effect, Pianka believes Australia also has a greater environmental heterogeneity—in other words, a larger R in Eq. (1).

The general conclusion of this section is that there are areas too small, and areas too large, to show clear diversity patterns, but that for the proper intermediate census area, the patterns are clear. Comparisons of pairs of these small census areas reveal the average multiple M, by which their combined census exceeds the separate ones. By using this, we can reconstruct how many species compound environments would have.

The Effect of Predators

If abundant predators prevent any species from becoming common, the entire picture changes. Resources are no longer of any concern and our Eqs. (1) and (2) are irrelevant. More correctly, resources are still a concern, but their manner of subdivision is irrelevant. J. Connell (pers. comm.) and Janzen (1970) have both shown that tropical tree seedlings seldom grow under their parent tree. This is no accident of seed fall;

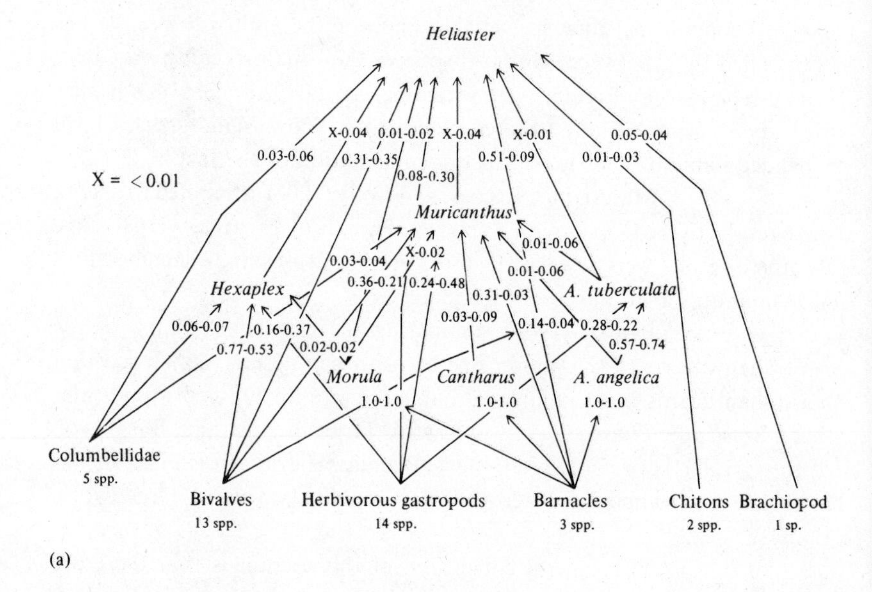

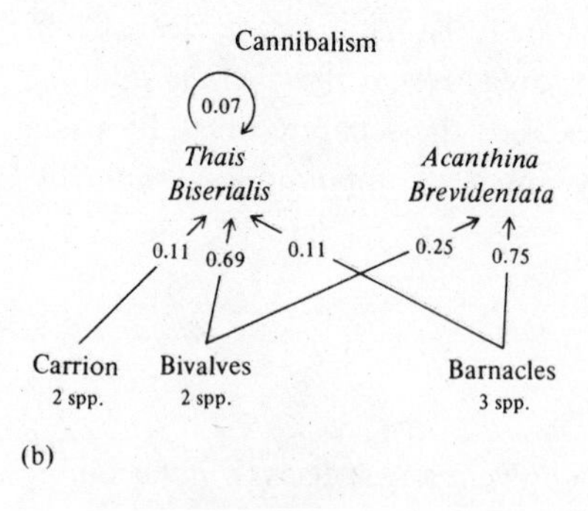

Fig. 7-9

(a) The complicated food web of a predator-rich community in the Gulf of California (genus A is *Acanthina*). Numbers on the left are fractions of numbers of food items in the predator's diet; numbers on the right are fractions of calories in the predator's diet. (b) A simpler food web in Costa Rica without a secondary carnivore. Numbers are fractions of numbers in the predator's diet. (From Paine, 1966.)

most must fall under the parent. But predators seem to gather where the seeds are commonest and there eat them. The result is that each kind of seedling grows more often under other species of canopy trees than under its own species. In itself, this does not allow a great tree species diversity, but a slight modification would. What we need is a ceiling on the abundance of each species. If predators gathered to collect the seeds only of common trees they would produce such a ceiling; any tree commoner than some ceiling abundance would be reduced by the predators. Alternately, if root parasites, fungi, or other diseases were acting as predators, nearby trees of the same species would be eliminated. Only isolated trees, far from sources of contagion, would be safe. In this way, too, an upper limit to the species' abundance would be set. Now, if we have ten tree species, each with maximum abundance of 1 per hectare, their canopies simply will not fill the space and there will be ground and sunlight for new trees to invade. They cannot be of the old species, so our community is vulnerable to invasion by new species. In this way there can be more species where predation is severe. This is result 3 on page 32.

This result is not automatic: Where predation is severe it is not inevitable that there will be more species. An indiscriminate predator would not focus on the common prey and reduce them to below a ceiling; neither would an indiscriminate predator preserve trees isolated from contagion. In fact Patrick (1970) has described how predation by the snail *Physa heterostropha* decreases diatom species diversity by selectively leaving the common diatom *Cocconeis placentula.* What is required is most easily achieved by species-specific predators, such as diseases and parasites. A predator with a "search image" that switches to whichever prey is commonest could put a ceiling on abundance and increase the number of species, but it will only have this effect if the ceiling it puts on total abundance is below the ceiling that resources would set. These ceilings are usually interacting: the predator ceiling is lower where resources are scarce because consumers are ill fed and poorer at escaping, and the resource ceiling is lower where predation is severe because it is more dangerous to search for food. What is relevant to an increased number of species is an independent predator ceiling low enough so that the community is vulnerable to invasion.

Paine (1966) has demonstrated another way that predation increases species diversity. When a starfish (*Pisaster*) feeds in an intertidal zone, it cleans a swath free from the mussels or barnacles that would otherwise outcompete the other species and dominate the community. In these swaths other species can maintain a sort of refuge. The intertidal community is then a mosaic of different succession stages. When Paine removed the predators, the number of other species fell from 15 to 8. Figure 7-9 shows how communities with more predators tend to be richer in species.

Predation certainly provides an alternative regulation of species diversity to that offered by competition. Probably the best evaluation is that both are sometimes important and that which one dominates depends upon the structure of the environment. Some environments are easy to search and are predator vulnerable; in these, predators doubtless exert a control over species diversity. In other areas prey species are hard to find and predators probably have little effect. Looking back over our results, the success of Eqs. (1) and (2) in environments differing in R and in stability is a suggestion that for these species groups, predation is not of importance.

Appendix 1 Derivation of Eq. (2)

Let the resource coordinate j and its utilization by species i, u_{ij}, be chosen so that $\frac{\sum_j u_{ij} u_{kj}}{\sum u_{ij}^2} = \alpha_{ik}$ (see p. 39). Let R_j be the total utilization by all species so that

$$R_j = \sum_i u_{ij} X_i \tag{3}$$

where X_i is the abundance of the ith species. Squaring both sides and summing over j we get $\sum_j R_j^2 = \sum_j \left\{ \sum_i u_{ij} X_i \sum_k u_{kj} X_k \right\} =$ $\frac{\sum_{i,k} \left\{ \sum_j u_{ij} u_{kj} \right\} X_i X_k}{\sum_j u_{ij}^2} \cdot \sum_j u_{ij}^2$ where we must (for simplicity) assume $\sum_j u_{ij}^2$ is independent of i. We will also assume $\sum_j u_{ij}$ independent of i; in combination these assumptions mean different species have the same U-shaped curves just displaced from one another. Then we get $\sum_j R_j^2 = \sum_j u_{ij}^2 \sum_{i,j} \alpha_{ik} X_i X_k = \sum_j u_{ij}^2 \left\{ \frac{\sum_{i,k} \alpha_{ik} X_i X_k}{\sum X_i^2} \right\} \cdot \sum X_i^2$. The term in braces is the "Rayleigh ratio," λ. Now we can relate the remaining terms to diversities, by noting, for instance, that $D_S = \frac{1}{\sum_i \left(\frac{X_i}{\sum X_i} \right)^2} =$ $\frac{\left(\sum_i X_i \right)^2}{\sum_i X_i^2}$ so that $\sum_i X_i^2 = \frac{\left(\sum_i X_i \right)^2}{D_S}$. Doing the same for other such expressions we find $\frac{\left(\sum_i R_j^2 \right)}{D_R} = \frac{\lambda \left(\sum_j u_{ij} \right)^2}{D_U} \cdot \frac{\left(\sum_i X_i \right)^2}{D_S}$. On summing Eq. (3) over j and then squaring, we find that $\left(\sum_j R_j \right)^2 = \left[\sum_i \left(\sum_j u_{ij} \right) X_i \right]^2 =$

$\left(\sum_j u_{ij}\right)^2 \left(\sum_i X_i\right)^2$ and the numerator squares cancel, leaving us, upon rearrangement, $D_S = \frac{D_R}{D_U} \cdot \lambda$, which is Eq. (2). The theory of Rayleigh ratios is quite subtle (e.g., Franklin, 1968) but we note that if all X_i are equal, λ directly equals $1 + \frac{1}{N} \sum_{i \neq k} \alpha_{ik} = 1 + C\bar{\alpha}$ where C measures the number of neighboring species (i.e., the number with which there is competition).

Appendix 2
Measures of Diversity

Numbers alone do not make science; it is relations between numbers that are needed. Applying a formula and calculating a "species diversity" from a census does not reveal very much; only by relating this diversity to something else—something about the environment perhaps—does it become science. Hence there is no intrinsic virtue in any particular diversity measure except insofar as it leads to clear relations.

Perhaps the word "diversity" like many of the words in the early vocabulary of ecologists ("sere," "ecotone," and others) should be eliminated from our vocabularies as doing more harm than good. To some people "diversity" means the number of species, to some it incorporates both the number and the evenness of their abundances, and to some it can be viewed as a vector with one component the number of species and the second component the evenness of abundances; to yet others it is best described by a relative abundance curve. Those who have used diversity to mean a number combining aspects of both the number of species and the evenness of their abundances have wasted a great deal of time in polemics about whether $\frac{1}{\sum_i p_i^2}$ or $-\sum_i p_i \log_e p_i$ or $\frac{N!}{N_1!N_2!\cdots N_n!}$ or some other measure is "best."

Here we have sometimes used only the number of species as a measure of diversity; sometimes we have used a whole function, and sometimes $\frac{1}{\sum_i p_i^2}$, incorporating number and relative abundance of species. $\frac{1}{\sum_i p_i^2}$ was used in preference to other, similar measures, such as $-\sum_i p_i \log_e p_i$, largely because of Eq. (2), which shows that, in the relation between this measure of species diversity and the same measure of production and utilization diversities, the multiplier that converts one to the other is related to the eigenvalues of the α matrix and has a biological interpretation. The corresponding relation between $-\sum_i p_i \ln p_i$ measures would be very complicated and not easy to interpret. (A purely subsidiary virtue of $\frac{1}{\sum_i p_i^2}$ is that it is very easy to calculate on even simple desk top calculators.) Thus, in competitive communities we have some reason to expect $\frac{1}{\sum_i p_i^2}$ to be a useful measure. But since not all communities need be competitive, we must consider other possibilities: there may not exist clear-cut diversity relations in noncompetitive communities, or there may exist clear-cut relations but only when a completely different measure is used. And, finally, even in a competitive community the number of species might, of course, be determined independently from the evenness, so that two or more relations would be involved rather than one.

Comparisons of Temperate and Tropics

8

A few decades ago it was fashionable for ecologists to study communities in the arctic on the grounds that these would be very simple communities and hence easy to understand. Many excellent ecologists still follow this belief, but there are others who feel that it may be easier to understand the extremely complex communities of the tropics. This sounds paradoxical: How can a more complex community be easier to understand? A possible answer might be that the complex community has strong interactions among species so that the lives of the separate species are less independent than in a simple community. Where there is greater interdependence, patterns may be more conspicuous.

Virtually every naturalist wants to visit the tropics. The wet tropical lowlands represent one extreme of nearly every spectrum of life. Most particularly they are rich in diversity of species, in diversity of structure, and in their general aspect of luxuriance. The wet tropical lowlands are not all the tropics have to offer, however. Tropical mountains and dry tropical lowlands are fascinating, and they too cast light on tropical patterns. Tropical oceans and particularly coral reefs have attracted naturalists as much as their mainland counterparts.

Since nearly every pattern of biogeography has its tropical aspects, this chapter overlaps with all the others. On the other hand, many fascinating features of the tropics appear at present not to exhibit biogeographic patterns, so we make no mention of them at all, and some are best treated wholly in other chapters. But many patterns of tropical-temperate variation are closely tied together, and their unity would be lost to view if they were not given special treatment. This chapter, therefore, presents those patterns that are best handled in a unified fashion.

Climatic Seasonality

What really marks the tropics is of course their climates. In the climate chapter we noted the general absence of protracted storms

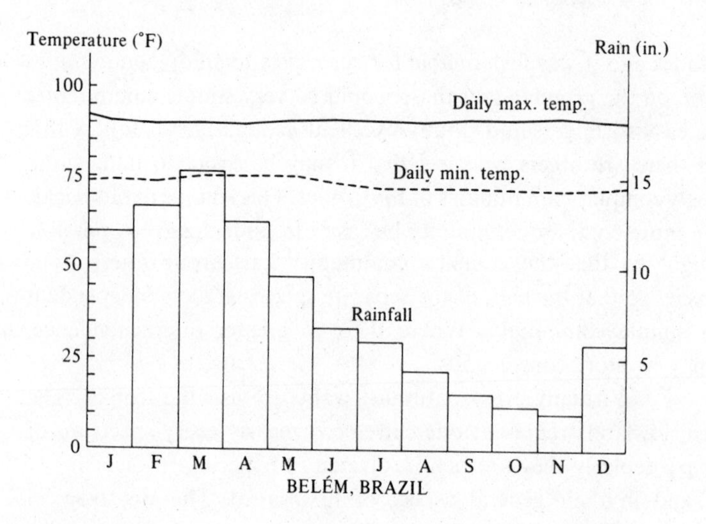

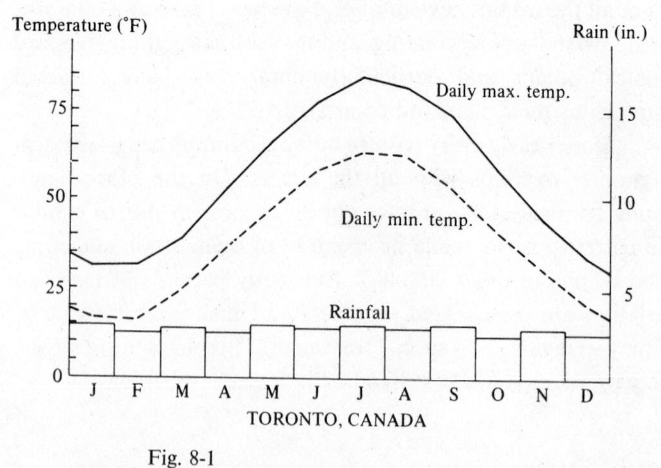

Fig. 8-1

Temperature and rainfall patterns in a tropical wet location, Belém, Brazil, and a temperate location, Toronto, Canada, showing that tropical areas tend to be seasonal in rainfall while temperate areas are seasonal in temperature.

in the tropics and, of course, their warmth due to their having the greatest amount of perpendicular radiation from the sun.

A comparison of the temperature and rainfall of Belém, Brazil, on the equator with Toronto, Canada (44°N latitude) (Fig. 8-1), will make a most important distinction. Basically, the wet tropics are seasonal in rainfall and the temperate regions are seasonal in temperature. So uniform are the mean monthly temperatures in most tropical areas that the average daily temperature range (daily maximum–daily minimum) exceeds the seasonal range (mean daily temperature of hottest month–mean daily temperature of coolest month). When things turn off for a bad season in the tropics, it is for rainfall reasons, not for temperature reasons. In fact winter never comes in tropical lowlands. Although there are temperate deserts with a pronounced dry season, the temperature change is still the change with biological consequences. Thus one or two bird species (e.g., phainopepla, *Phainopepla nitens*) move about in Arizona apparently to avoid the dry season, whereas about 100 species leave to avoid winter cold. We note also that no bird breeding in Colombia (Schauensee, 1964) or Surinam (Haverschmidt, 1968) leaves those tropical countries to avoid an off season. Thus birds find a temperate winter more difficult to face than a tropical dry season. Many trees do drop their leaves in the dry season, and in more arid tropical regions the effect can be almost like winter. Thus both tropics and temperate have seasons of a kind. We have seen that increased seasonality is one of the major components in species diversity patterns, so it would be useful to know whether the tropics or temperate seasons are more drastic. In spite of the underrated importance of tropical dry seasons, there can be no question that temperate zone seasons are more drastic for most species.

Temperate zone trees produce annual growth rings, reflecting rapid summer growth contrasted with very slow winter growth. In wet tropical lowlands the trees do not produce conspicuous rings. This poses a problem for the ecologist who wants to know age-specific birth and death rates for tropical trees, since there is no simple way to determine age from growth rings, but to us it simply means that tropical tree growth is more uniform throughout the year than tree growth in the temperate zone. Not only is growth more uniform, but production of fruit is less seasonal. Some tree species are fruiting at any season, making fruits and berries a feasible diet in the tropics (Figs. 8-2 and 8-3). Thus

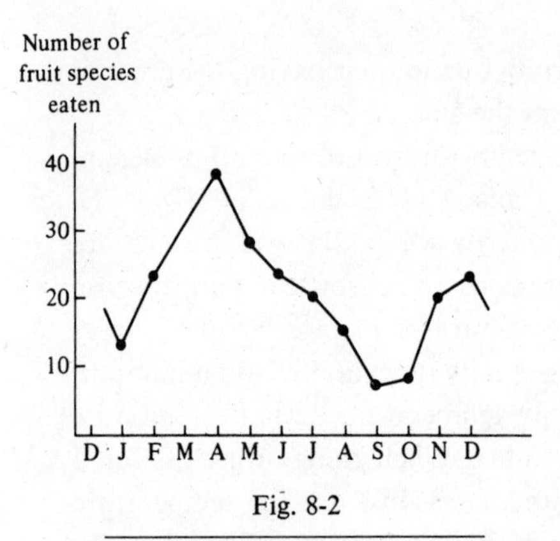

Fig. 8-2

Number of fruit species eaten by manakins (*Manacus manacus*) in Trinidad each month. (After Snow, 1962.)

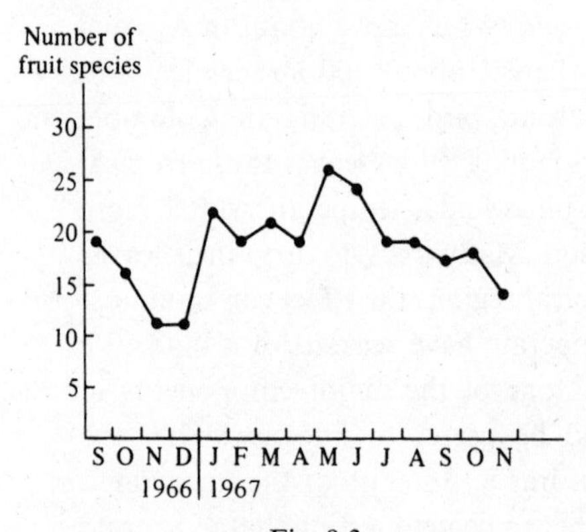

Fig. 8-3

Number of fruit species falling into containers in the forest of Panama each month. (From Smythe, 1970.)

there seem to be more fruit-eating bats than insect-eating ones in the tropics, whereas all are insectivorous in the North American temperate. No temperate zone bird or mammal can afford to specialize on fruits, although many generalists, like American robins, are happy to eat small fruits in their season, turning to other foods when fruit gives out. Of course different kinds of fruit have characteristic seasons even in the tropics, so an insect that eats only one fruit species must have a life cycle that is seasonal. But different insects eat the other fruits that ripen at other seasons. For this and other reasons some kinds of insects are available at all seasons in the tropics. Consequently bird breeding seasons are greatly elongated in the tropics, reflecting reduced seasonality. All of these lines of evidence show that tropical organisms react to their environment as less seasonal than do those in the temperate. Of course a climatologist can concoct an infinite number of measures of seasonality, some of which would show the tropics more seasonal. But we are interested in the organism's view, not the scientist's, so we say the tropics are less seasonal.

It has been claimed by naturalists, experienced in both the temperate and the tropics, that yearly fluctuations in animal populations are even more pronounced in the tropics than in the temperate, and this has been regarded as biological evidence for greater seasonality in the tropics. But this kind of argument is slipshod. Population fluctuations are larger not only in a more seasonal climate but also where competitors are packed more closely. This is just a restatement of one of our competition conclusions: that closely packed species with large α's have more precarious coexistence. In view of the foregoing evidence of reduced seasonality in the tropics, we tentatively infer that any great tropical population fluctuations are the consequence of closely packed competitors.

There is one indirect effect of temperate seasons that must be of considerable importance in the tropics. Temperate temperature seasons force vast numbers of temperate breeding birds to winter in the tropics. Hence there is a large winter influx of temperate zone birds. The scale of this influx can be impressive. Most North American tropical migrants winter in the small area of Central America, the West Indies, and northwestern South America. Few go as far as the vast forests of Brazil and Peru. Nearly half of the breeding birds of the great extent of coniferous forests in Canada and more than half of the birds of eastern deciduous forests winter in these tropics (Fig. 8-4). This area supplying the tropics is well over 1 million square miles in extent, with an average bird population of at least 2 bird families (pairs plus their offspring) per acre or 1280 families per square mile. If half go to the tropics, then at the very minimum 640 million families of birds are added to the resident birds of the small tropical wintering grounds each winter. By these calculations, the tropics in winter should be seething with migrant birds. Migrants are certainly conspicuous in tropical second growth and on mountains, but not in the numbers we would expect. And the rain forests hold relatively few migrants, presumably because they are more packed with competitors. Where most of the migrants go is really a mystery, but there is no doubt that those that do appear in the tropics must exert a profound influence on the food supply. In this sense there is a food shortage in the winter wherever there are many migrants, and its main effect seems to be to delay the breeding of native counterparts of the migrants until after the temperate zone birds have departed.

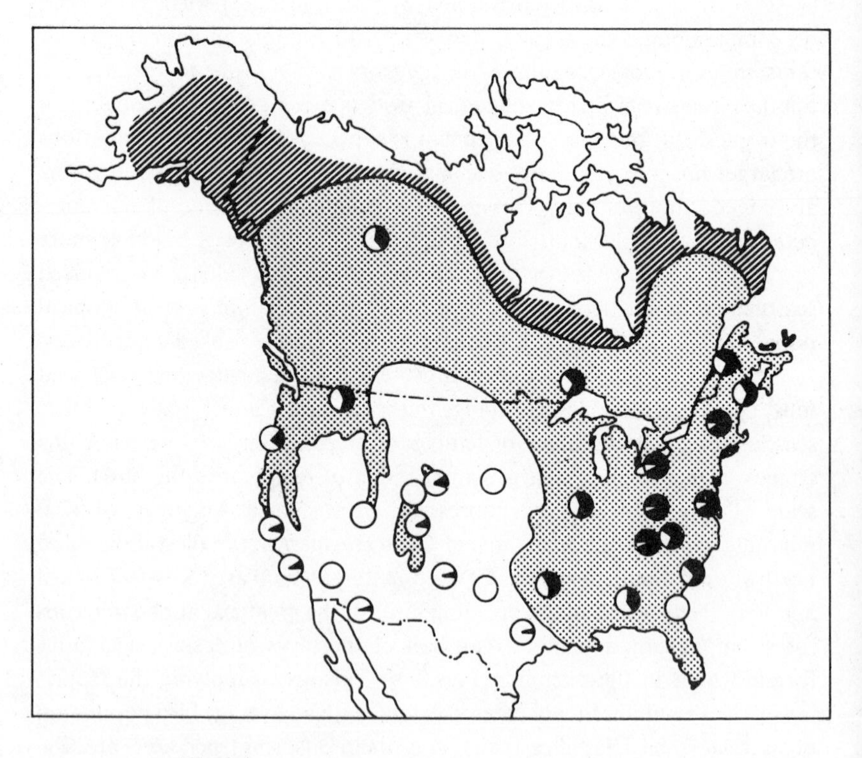

Fig. 8-4

Proportion of neotropical migrant individuals. The black section of the circles represents the proportion of breeding bird individuals in undisturbed vegetation communities at the locality that will migrate out of the Nearctic region in the winter. The stippled zone is roughly the forested region. (From MacArthur, 1959.)

The Spectrum of Resources in the Tropics

Since tropical and temperate birds are well known, it is natural to try to assemble comparative information on the spectrum of their resources in temperate and tropics. As we have already seen (Chapter 7), bird resources are subdivided in several dimensions. Different birds feed at different heights above the ground and in different foliage densities, on perches of different diameters and on foods of different sizes. Furthermore, some pick stationary insects from foliage, either while standing perched or while hovering on the wing; some eat fruit; some flycatch and others search tree trunk and bark crevices; some feed in flowers from nectar and associated insects. Each of these suggests a dimension of the bird resource spectrum; it should have a coordinate for food size, for food height and foliage density, for food mobility, and for other less easily plotted dimensions.

Thanks to Janzen and to Schoener (1971), we actually know more about the distribution of insect sizes in the tropics than in the temperate. However, we know enough about both to illustrate the contrast. Figure 8-5 compares the distribution of insect sizes, as gathered by sweep net samples, in the tropics and in one temperate area. Not only is the mean insect size less in the temperate forest, but the variance in size is also less; that is, in the tropics there are more large and as many very small insects, so the tropical range of insect sizes is greater. This should of course cause a greater diversity in the sizes of birds eating the insects. Better, since the bird's bill is the tool with which it catches insects, we expect a greater range and diversity of bird bill lengths in the tropics. Schoener (1971) has shown this to be the case (Fig. 8-6). In Fig. 8-6 only insectivorous birds are included and we see that tropical birds indeed have a wider range of bill lengths, presumably to harvest the wider range of insect sizes. We now have documented one resource coordinate that is not only wider in the tropics but also more widely utilized. What about the other coordinates?

It would seem that the range of feeding heights and foliage densities is about the same in the tropics as in the temperate. Forests are occasionally taller in the tropics than all but the American west coast conifer and Australian *Eucalyptus* forests in the temperate, but many tropical forests seem to be no taller and to have no greater range of foliage densities. In fact bushes, which are very dense low foliage, have a configuration seemingly absent from the understory of rain forests. Here, then, is a resource coordinate that is no longer and is perhaps shorter in the tropics. It is, however, subdivided more finely by the birds, but that story is told in a later section of the chapter.

Looking at other bird resource coordinates it is not clear that any except food size is longer in the tropics, but there are some entirely new dimensions that appear. Owing to reduced seasonality we have seen that fruit is perennially available in the tropics so that an insect → fruit dimension is possible. Also, there are structures in tropical trees not even present in the temperate; the great pineapple-like bromeliads that abound on the large tree limbs harbor many invertebrates and grow flowers providing a completely novel food source.

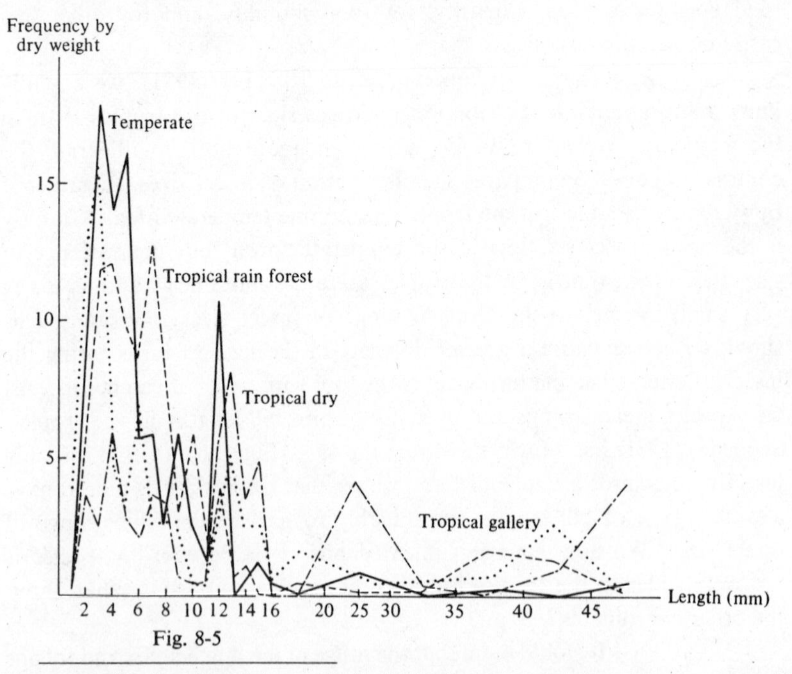

Fig. 8-5

Frequency distributions of dry weight by length for arthropods from sweep samples of tropical and temperate forest understories. Total dry weights for samples in milligrams: temperate, 3444; tropical dry, 3365; tropical gallery, 5344; tropical rain forest, 2661. (From Schoener, 1971.)

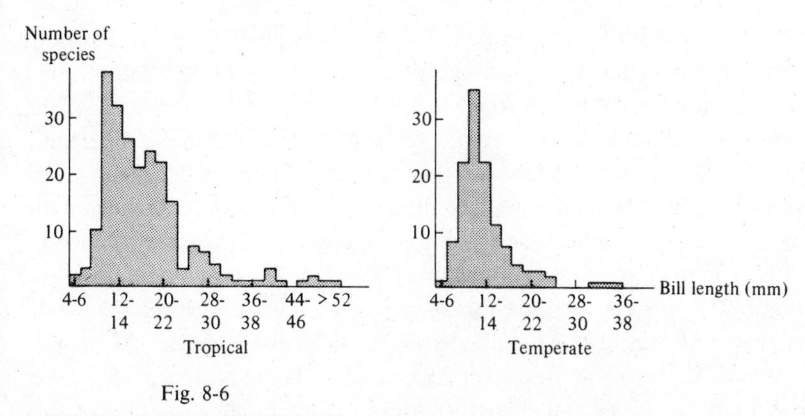

Fig. 8-6

Distributions of bill lengths for insectivorous (70% or more arthropod food) species of birds breeding in a tropical (8°–10°N) and a temperate (42°–44°N) latitudinal transect. (From Schoener, 1971.)

For organisms other than birds the picture is not always clear. For monophagous insects, however, it is very clear that the tropical spectrum of resources is longer. These species each eat a single species of plant, and since there are many times more plant species in the tropics than in most temperate areas, there is a much wider range of food for insects. The epiphytic bromeliads also provide new ways of life for insects.

We conclude that not only are the tropics less seasonal but they often offer a wider range of resources in a more highly structured habitat.

Patchy Distributions in the Tropics

How do life histories and species geographic ranges differ in these less seasonal and more highly structured tropics? Again, our information must come largely from birds, where mortality rates are known, clutch sizes are observed, and bird watchers have been noting geographic distributions for a century.

In the first place tropical species often have patchy geographic distributions for reasons that have no obvious relation to climate or habitat. In the rich fauna of New Guinea, this has been described for both ants and birds. Wilson (1958) studied and described three localities within 12 kilometers of one another in a forest region of New Guinea that had not been severely disturbed by man. Each area contained all of the major ant habitat types of the region and yet there were marked differences in distribution among the species. A species of the genus *Crematogaster* was moderately abundant in the Didiman Creek forest but absent from the other two, and another species of the same genus was among the two or three commonest species at Didiman Creek and was absent from the Buser River forest. At least three ant species were moderately abundant at Buser River and missing altogether from the Bubia forest. Of the two commonest species at Bubia, one was rare at Didiman Creek, the other rare at Buser River. All these disparities existed in spite of the fact that each locality had every habitat used by the ant species.

Diamond (in press) has pointed out that the New Guinea birds offer many examples of species missing for reasons that do not seem to be climatic or due to poor habitat. Figure 8-7 shows the range of the Papuan creeper (*Climacteris placens*); the blank area in the eastern highlands seems in every way suitable and yet the species is missing. Since it is

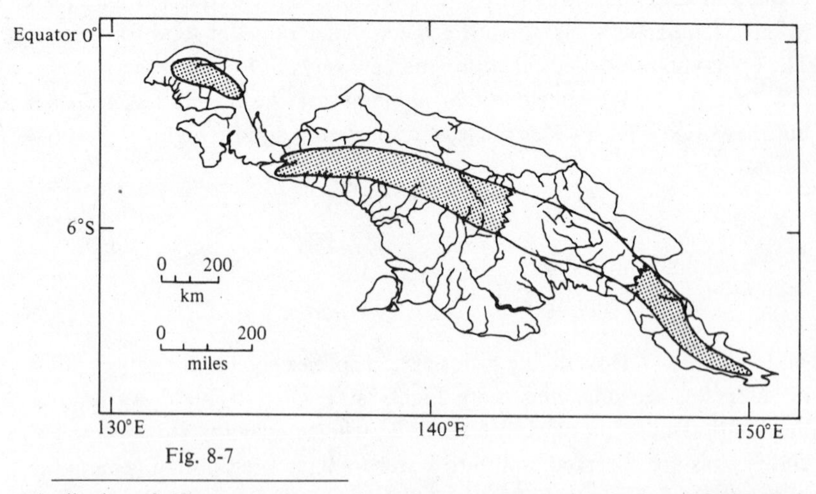

Fig. 8-7

Distribution of *Climacteris placens*, the Papuan creeper, in New Guinea. This species is present on the Vogelkop and on the western and southeastern portions of the central range but is absent from the eastern highlands, so that its distribution is discontinuous despite the continuous expanse of montane habitat on the central range. (After Diamond, in press.)

present on both sides of the blank area, we would find it hard to claim that the result was historical and that the species had never tried to colonize the eastern highlands. Diamond goes to some lengths to show that searches have been careful enough so that the species would have been observed had it been present, and he points out that there are 8 other such species that drop out in the eastern highlands.

New World tropical birds offer abundant examples along the same lines. They are doubtless commonest in South America, but since that continent has not been thoroughly explored, the skeptic may claim that gaps in distributions are due to lack of careful search. Central America has been more thoroughly studied and many distribution gaps in that area cannot be explained away as the result of lack of observers. Panama, especially the Canal Zone region, has been more carefully studied than any other part of Central America. Yet Eisenmann (1955) notes several species found along both sides of the country but absent from it even though Panama has an abundance of apparently suitable habitat. Such species include common ground dove (*Columbina passerina*), green jay (*Cyanocorax yncas*), and vermilion flycatcher (*Pyrocephalus rubinus*), found north to the United States and also in South America. Many more Central American birds show a more local kind of patchiness in their distributions—abundant in one place and very rare nearby. For example, the white-fronted nunbird (*Monasa morphoeus*) is common and conspicuous in Caribbean Costa Rica and certainly very rare, if not presently absent, from the same slope of central Panama.

Do temperate zone species have such patchy distributions? A few have fragmented distributions due to human persecution. The wild turkey (*Meleagris gallopavo*) is a typical example. Far more species have fragmented distributions because their habitat is fragmented. Birds confined to mountain vegetation are the commonest cases of this kind. But only three species of birds of United States or Canada have mysterious gaps in their ranges as shown on the maps in Robbins, Bruun, and Zim (1966). Two of these are the black swift (*Cypseloides niger*), whose cliff nesting sites may in fact be of a type which is fragmented; and the northern three-toed woodpecker (*Picoides tridactylus*), which is apparently missing from the conifer forests of north-central Canada although found both east and west of there and though its close congener, the black-backed three-toed woodpecker (*P. arcticus*), has a continuous range. Conceivably this gap is due to insufficient observation; rather more people look for birds in Panama than in northern Saskatchewan! The third is the cardinal (*Cardinalis cardinalis*), which is missing from deserts and rivers of west Texas and eastern New Mexico although it occupies those habitats in Arizona and is found throughout eastern United States. It is probably no accident that the pyrruloxia (*Cardinalis sinuata*) is found primarily in the gap of the cardinal range although they are sometimes found together in Arizona. In summary, there seem to be very few range gaps found among American birds of the kind that seems quite common in the tropics, and we conclude tropical birds have patchier distributions.

Essentially four explanations have been proposed for these patchy distributions: (1) In history the gap areas contained no suitable habitats and only recently have such habitats existed. The species have not yet had time to recolonize. (2) Competition, usually diffuse competition, prevents the species from persisting in the gap areas. (3) The habitats in the gap areas are not really suitable; only our ignorance makes us think the species' habitats are continuous. (4) The patches are remnants of a once continuous distribution of a species going extinct. Explanation (3) is not really independent of the competition explanation, (2). The more competitors there are, the subtler the properties of the habitat where a species can persist. Hence (2) and (3) may be the same explanation. Even the historical explanation, (1), could interact with competition. In the absence of competitors recolonization may be much more rapid than in their presence. However, it would be hard to claim that the tropics have had a more disturbed last million years than the temperate zones that were invaded at least four times by continental glaciers. Hence the likely explanation of the greater tropical patchiness seems to be that competitors are packed more closely, causing a species to persist only where very subtle and particular conditions are met. See the chapter appendix for further discussion of tight packing of species and patchy geographic distributions.

Tropical Richness of Species

One of the most conspicuous facts about the tropics is their diversity of life. In many areas there are well over 100 tree species, many so rare that there may be only one specimen in 100 hectares. The numbers of insect species are not known; often the species have never been named, but naturalists are aware of a great tropical diversity. Schoener and Janzen (1968), with 2000 sweeps of a net, collected 545 species in a wet lowland tropical forest. Two thousand sweeps in a mixed Massachusetts forest yielded between 360 and 410 species for each month. Darlington showed the number of ant species at various south latitudes in South America (Table 8-1). The insects not only come in many species but also in many sizes and forms. The gradient in numbers of bird species is best illustrated by Fig. 8-8, which shows how the numbers of breeding land bird species grow toward the tropics. There is every reason to think that most other terrestrial taxa of large enough size show similar patterns. Thus there is a great tropical increase in amphibians, although salamanders as a subgroup of amphibians may be richer in the temperate. Marine faunas too seem to be richer in the tropics. Certainly coral reefs are famous for their richness in species and in their forms. Figure 8-9 shows the number of mollusc species at various stations along the Atlantic coast, following Fischer (1960). Figure 8-10 shows the pattern for planktonic Foraminifera following Stehli (1968). Certainly tropical richness in species is a very common phenomenon.

There are two exceptions to great tropical diversities that are intriguing. First, tropical mountaintops have fewer species than tropical lowlands. We saw a detailed example of this in Fig. 5-19, where we explained it as partially an island effect. But even in continuous mountain chains, the highlands are impoverished and tend to be occupied by a rather temperate-like fauna. Figure 8-11 from Kikkawa and Williams (1971) shows data for the whole of New Guinea. To give another instance,

Table 8-1
Decrease in Number of Species of Ants Southward in South America

Area	Approximate S latitude	Number of species
São Paulo, Brazil	20°–25°	222
Misiones, Argentina	26°–28°	191
Tucumán, Argentina	26°–28°	139
Buenos Aires, Argentina	33°–39°	103
Patagonia as a whole	39°–52°	59
Patagonia, humid west	40°–52°	19
Tierra del Fuego	43°–55°	2

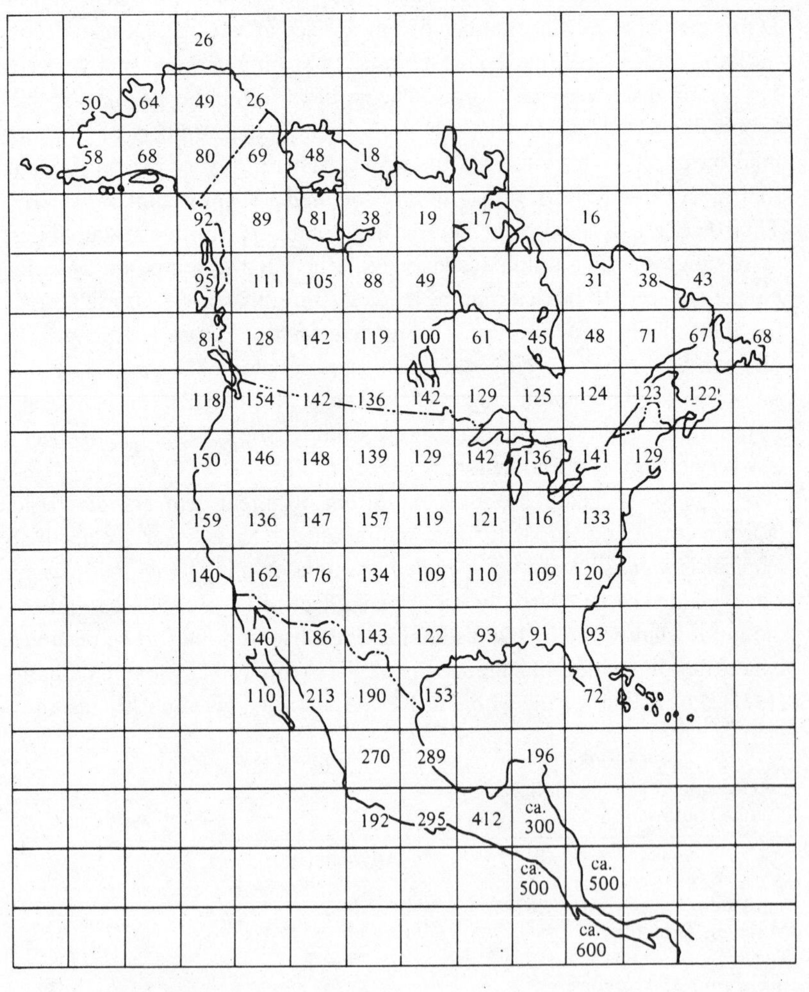

Fig. 8-8

Number of breeding land bird species in different parts of North America, from various sources. (From MacArthur, 1969, after MacArthur and Wilson.)

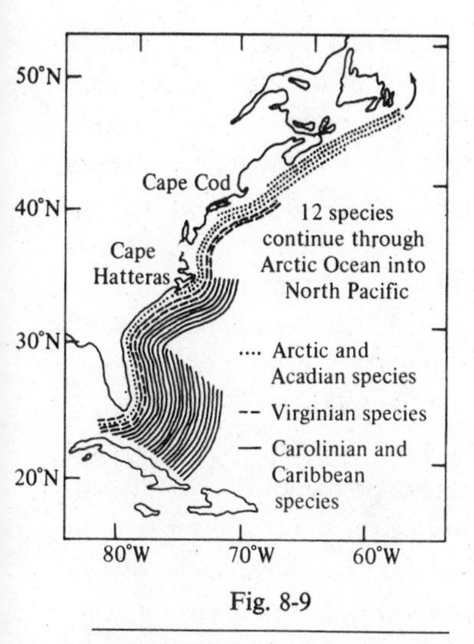

Fig. 8-9

Diversity gradient of gastropods along the eastern coast of the United States and Canada. Each line stands for ten species. (From Fischer, 1960, after Abbott.)

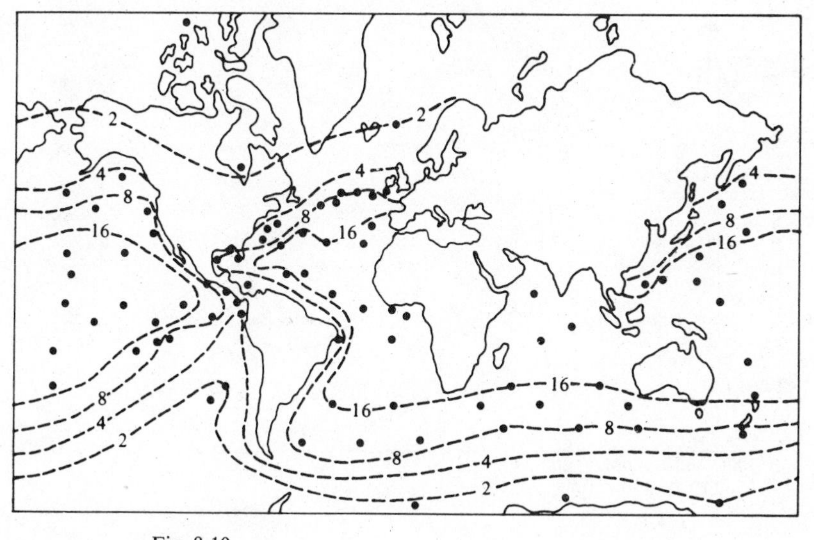

Fig. 8-10

Contoured raw diversity for Recent species of planktonic Foraminifera. (From Stehli, 1968.)

Central American mountains have lost, or nearly lost, many typical tropical lowland bird families such as antbirds (Formicariidae), puffbirds (Bucconidae), and manakins (Pipridae) and tend instead to be relatively rich in a few species of warblers (Parulidae), vireos (Vireonidae), finches (Fringillidae), and hummingbirds (Trochilidae), all of which are prominent in the temperate zones. But the mountains are not replicas of temperate zones! Although the mean annual temperature and the short height of the trees may be temperate, there is no winter. In fact seasonal temperate variation may be less on the tropical mountains than in the lowlands, due to a perpetual cloud of moisture. However, it does freeze on the mountains, and insects that dislike freezing weather have no summer, when they can count on no frosts, or winter, when the frosts are predictably concentrated. Productivity is doubtless lower and the range of resources shorter than in the lowlands. And even wet tropical mountains may have spells of dry weather when the wind makes everything crisp with lack of moisture. This gives a kind of seasonality to tropical mountains, but not an exact counterpart of the temperate seasonality. In this environment there are few species.

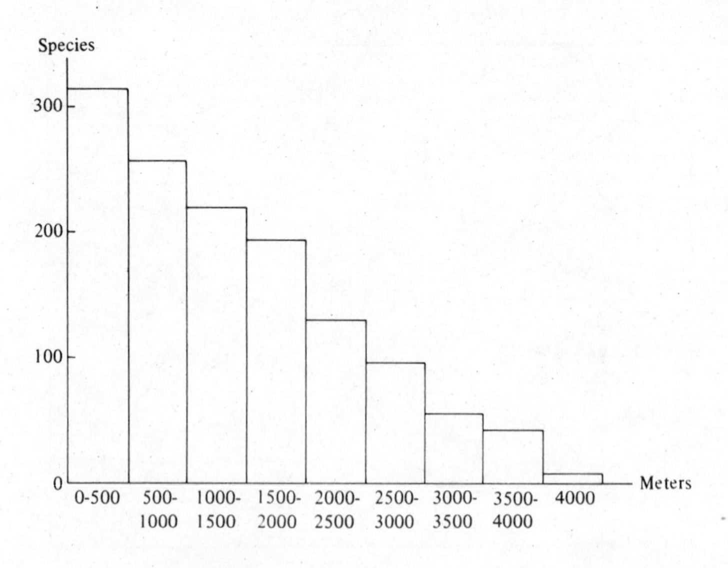

Fig. 8-11

Number of bird species in each altitudinal zone on New Guinea. (From Kikkawa and Williams, 1971.)

The other chief exception to the rule of increased tropical diversity occurs in freshwater. Patrick (1966) compared the numbers of species of major groups of organisms found in hard waters of the upper Amazon basin (Rio Tulumayo and Quebrada de Puente Perez) with waters of comparable chemistry in temperate North America (Ottawa River and Potomac River). Table 8-2 from Patrick shows that the temperate waters are fully as rich as those of the Amazon basin. The tropical waters were not more productive than the temperate ones and were perhaps even more seasonal because the seasonal tropical rains caused periodic flooding of the rivers.

We saw in Chapter 7, on species diversity, that it is possible for one large area to have twice as many species as another without any of its component habitats having more species. The explanation would lie in a greater faunal difference between habitats in the richer area. But this is not the sole explanation of high tropical diversities as is shown in Fig. 8-12. The temperate data are typical of temperate forests nearly everywhere in the United States; the tropical data are probably moderately reliable for large areas although a few new species will be added with further observation. The small tropical areas are from Barro Colorado Island in Gatun Lake in the Canal Zone. Gatun Lake was made when the Panama Canal was constructed, so Barro Colorado Island is well under a century old. Even so, it seems to be poor in species compared to the adjacent mainland, and the 5-acre census may also be an underestimate. In any case, even small tropical areas of 5 acres are at least two

Table 8-2
Comparison of Species Numbers in Temperate and Tropical Waters
(From Patrick, 1966)

	Río Tulumayo	Quebrada de Puente Perez	Ottawa River 1955–1956	Potomac River
Algae	73[a]	62	69	103
Protozoa	33	40	47	68
Lower invertebrates[b]	5	6	8–15	27
Insects	104	78	51–64	104
Fish	26	22	17–28	28

The first two rivers are in Peru; the second two, in North America.
[a] These numbers represent established numbers of taxa and do not include taxa represented by less than six specimens.
[b] Excluding rotifers.

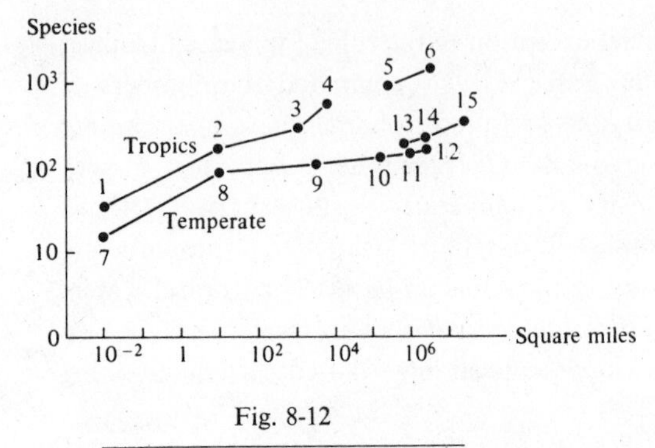

Fig. 8-12

Number of breeding land bird species plotted against the area in square miles for tropical and temperate areas. 1, 5-acre census on Barro Colorado Island; 2, Barro Colorado Island (6 sq miles); 3, Panama Canal Zone; 4, Republic of Panama; 5, Ecuador; 6, Colombia plus Ecuador plus Peru; 7, 5-acre Vermont census; 8, 6 sq miles in southern Vermont; 9, southern Vermont; 10, New England; 11, northeastern United States and adjacent Canada; 12, eastern United States and Canada; 13, Texas; 14, western United States; 15, North America (north of Mexico). The points for western United States indicate the effect of greater topographic diversity. (From various sources, after MacArthur, 1969.)

and one-half times as rich in bird species as temperate areas. These species live so close together that they are coexisting by some device. In the terminology of the last chapter, either the spectrum of resources, R, is greater in the tropics; the utilization per species, U, is less in the tropics; the overlap between species, $\frac{\bar{O}}{\overline{H}}$, is greater in the tropics; or the dimensionality of the environment, C, is greater in the tropics. Any of these can cause coexistence of the increased number of species that has been observed in the tropics. The embarrassment is that all are likely to be true, at least in some places! We have already seen that R is greater and probably also C. The reduced seasonality will allow a smaller U (p. 202) and larger overlap (Chapter 2, and appendix to this chapter); and the greater patchiness of tropical species distributions is evidence that overlap really is greater.

The data of Sanders (1968) are of considerable interest in this context. (His methods were discussed in detail on pages 180–182, to which the reader can turn for interpretation of the graphs.) Figure 8-13 shows Sanders' comparisons of tropical and temperate marine benthic polychaete and bivalve diversities. Sanders as before interprets these patterns as consequences of reduced tropical seasonality. The Bay of Bengal site was too deep to be affected by seasonal, monsoon freshening of the water and he inferred that it was very stable and hence diverse. This certainly seems a reasonable inference but as before we may question whether other factors also influence the diversity of his samples. Do we know whether there is an increased spectrum of resources in the tropical benthic sediments? What about productivity? and structural complexity? Lacking this knowledge we are not in a position to infer that the tropical climatic stability is the only factor, but it surely seems to be a primary one.

We have also seen that more intense predation can increase the diversity of prey species. Is predation more intense in the tropics? This is a complicated question. It appears that predation is more intense on nestling birds but perhaps not on adults. That it is very intense on nestling birds is the consensus of naturalists who have studied nesting

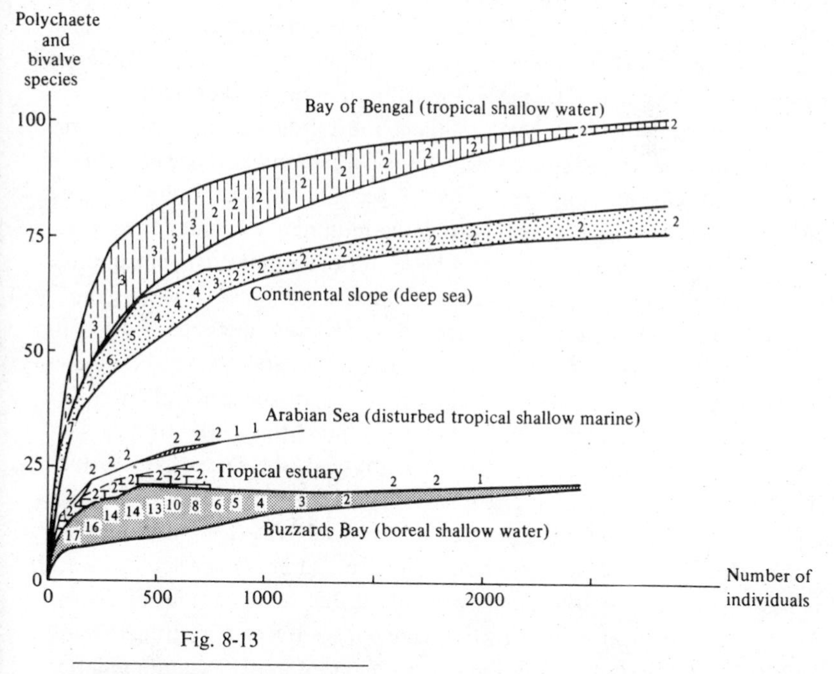

Fig. 8-13

Range for diversity values found for a number of regions in the temperate and the tropics. (From Sanders, 1968.)

success, or rather failure, in the tropics (e.g., Skutch, 1954, 1960, 1969). Data in which the same species is compared in the temperate and the tropics were kindly provided by G. Orians as follows:

	Costa Rica	Washington
Nests fledging at least one young	20	225
Nests fledging no young	73	274
Total nests	93	499

This doesn't quite give an estimate of mortality since the number of eggs per nest is less in the tropics, but predators usually ruin all of the eggs in a nest, so the nests from which no young fledged are usually those that were attacked by predators. We conclude that nest predation, at least for red-winged blackbirds (*Agelaius phoeniceus*), is more intense in the tropics. On the other hand adult mortality is far less. We can prove this for birds and can suggest why it is likely for many other organisms. Snow (1962) made an intensive study of the tropical black and white manakin (*Manacus manacus*) in Trinidad. He put colored aluminum bands on the legs of adult males and observed 89% of them the following year. Since some birds might have left the study area, 89% is a minimum estimate of survival. Karr (1971) banded many species over 2 years in Panama and recaptured similarly high percentages, again providing minimum estimates of survival. For the temperate we can provide a better estimate than that. The number of recaptures after 2 years divided by the number after 1 year is much closer to a true survival rate. Farner (1955) reviewed methods and results for temperate birds. For temperate passerine birds, such survival estimates averaged about 50%. These are far smaller than the tropical recapture rate. Hence there seems little doubt that adult birds in the tropics have much greater success in surviving than those in the temperate. In fact, the half-life of a bird with 0.9 survival is that number C such that $0.9^C = 0.5$. This $C = 6.6$, so half of the tropical birds are alive after 6.6 years. Half of the temperate birds are dead after somewhere between 1 and 2 years. These mortalities are not all due to predation, of course. Storms and the hazards of migration take a large toll in temperate regions. But the low adult mortality in the tropics makes it hard to claim predation is intense there, at least on adults. There seems to be no evidence that nest predators switch to whichever species is common and regulate populations in that way. This requirement for predator-controlled tropical species diversity in birds is therefore not met, and we tentatively discard this explanation. We only do so for birds, however; there is abundant evidence, as we mentioned, for trees and marine intertidal organisms. Janzen (1970) gives evidence that tropical tree species are kept sparse by herbivores who thus allow more species to fit in.

Finally, there is yet another explanation of tropical diversity that has been proposed by Fischer (1960) among others. According to this view the tropics had a head start in accumulating species either because more speciated there or fewer went extinct; given enough time the temperate species may increase until there are eventually as many in the temperate as the tropics have now. At first sight this explanation looks incompatible with the others; either the tropics are full or they are not. If they are full, history is irrelevant; if they are not full, competition is irrelevant. Now, for species of given morphology and physiology there may be such a thing as a full community (see chapter appendix 1), but "full" would apply only to small communities of coexisting species. The number of species in a large area of many habitats could easily keep increasing by increasing the faunal differences between habitats. It is significant that Fischer was describing the richness of whole faunas covering wide areas, so we have sufficient reason for history to have been important. Even if the component habitats are not full, however, history need not be the only relevant factor. We saw that a principle of equal opportunity for further colonization should prevail, producing a balance between the numbers of species in adjacent habitats. This balance is understandable in terms of competition even if, through history, both habitats are becoming enriched. The critical question, whether the number of species is increasing as the historical explanation would suggest, remains unsolved. Figure 8-14 from Simpson (1965) suggests the contrary, as does Brodkorb's (1971) estimate that the number of bird species has, if anything, declined since Miocene times.

Tropical Clutch Sizes

There is an interesting unexplained pattern in birth rates, or at least clutch sizes—the number of eggs per nest. Most small temperate

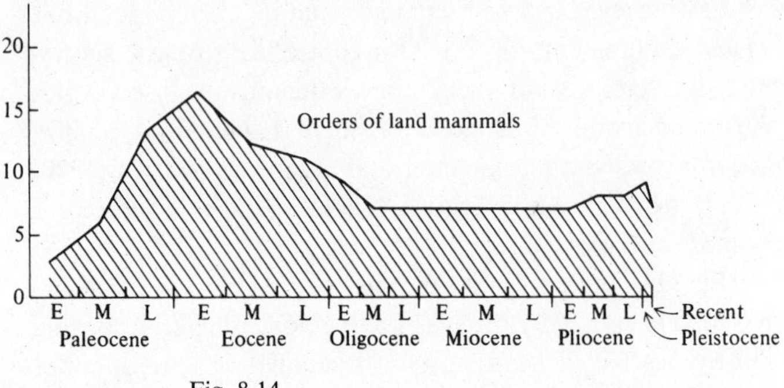

Fig. 8-14

Changes in fundamental diversity (number of orders) in the North American fauna. (From Simpson, 1965).

land birds lay 4–5 eggs to a clutch, whereas most small tropical land birds lay 2 eggs to a clutch (Table 8-3). Three subsidiary patterns complicate the main one. First, birds that nest in holes in tree trunks or banks have much larger clutches than is general—often twice as large. Second, birds that depend upon outbreaks of superabundant food also have larger clutches. Third, birds of tropical mountaintops, poor in species, have low clutch sizes, according to some information in Skutch (1967). Thus among

Table 8-3
Comparison of Clutch Sizes in the Temperate and the Tropics

Clutch size (median)	Number of land bird species: Western U.S.[a]	Central America[b]
1	3	1
2	25	81
3	43	33
4	131	9
5	73	4
6	24	
7	10	
8	4	
Mean	4.208	2.484
σ	1.240	0.773

[a] From Peterson (1961).
[b] From Skutch (1954, 1960, 1969).

North American wood warblers almost all temperate species lay about 4 eggs, but the hole-nesting prothonotary warbler (*Protonotaria citrina*) lays 6, and the Cape May (*Dendroica tigrina*) and bay-breasted (*D. castanea*), that depend upon outbreaks of spruce budworms (*Choristoneura fumiferana*) for their summer food, also lay large clutches of 5–7, or even more, eggs.

The fact that there is a great spring bloom of food in temperate zones, where clutches are large, coupled with the fact that species depending upon periodic outbreaks of food have even larger clutches, suggests that clutches are larger where parents can feed more young. But why do hole nesters have such large clutches?

It has been suggested that larger clutches attract disproportionate numbers of predators and that there are fewer predators both in the temperate and on hole nests, allowing these to have larger clutches. This explanation does not account for the large clutches of outbreak species, or for the fact that tropical islands, without many predators, have small clutches like the mainland.

Cody (1966) has attempted to amalgamate the separate hypotheses into one in which the need for compromise among conflicting demands on the parents is central. Where predators are important, the bird must devote more time to their avoidance and thus make some sacrifice in clutch size; where species have to recolonize frequently, as in the temperate regions, and especially in outbreak species, the clutch size must be large to give the founding populations a good start, in which case other qualities are subordinated. This kind of explanation is appealing because it provides so much leeway and can be adjusted to account for almost any pattern.

Moll and Legler (1971) have provided a very clear exception to the rule that species have smaller clutches in the tropics. Although most turtles, like birds, lay few eggs in the tropics, the slider turtle, *Pseudomys scripta*, not only lays far more eggs in Panama than other tropical turtles, it also lays more there than in temperate United States. Thus the average clutches from Illinois and Louisiana were 9 and 7 whereas the Panama average clutch was 17 (small eggs). The typical tropical turtle lays much smaller clutches of 1–3 large eggs. Legler has suggested that the increased clutch size in Panama may in part be due to an increase in turtle size. Unlike birds, large turtles lay more eggs than small ones, and the Panama slider turtles are very large. Moll and Legler regard the slider as a new arrival in the tropics—apparently it has not lost its temperate reproductive habits. Here is an opportunity to compare the virtues of the temperate and tropical strategies.

Some ingenious observations are needed to shed new light on this old problem.

Global Equilibrium of Species

Whether or not species are still accumulating through history, we can still apply the principle of equal opportunity to find the number of species at any place on earth. We suppose only that species adjust their distributions more rapidly than they speciate. Two areas A_1 and A_2 are at equilibrium, with equal opportunity, if an individual of species 1 from A_1 and an individual of species 2 from A_2 would do exactly equally well if they traded areas. We have already seen that if the production of resource j is the same for all j and for both habitats, equal opportunity means equal overlap in the two habitats. Of course these assumptions are patently false, but the result is approximately true even with differing productions. Equal overlap means equal $1 + (n - 1)\bar{\alpha}$. But Eq. (2) of Chapter 7 says the diversity of species $D_S = \frac{D_R}{D_U}(1 + (n - 1)\bar{\alpha})$ where D_R and D_U are the diversity of production and utilization. With the term in parentheses constant, equal to M, the diversity of species equals $M\frac{D_R}{D_U}$. In terms of Eq. (1) of Chapter 7, the number of species is $M\frac{R}{U}$, which is equivalent. In other words one habitat or latitude will have more species than another if and only if its $\frac{R}{U}$ is greater. Tropical latitudes have larger R as we have seen. They can also afford to have smaller U. For these reasons, even if the species are still accumulating through history, the tropics would be expected to hold many more species than higher latitudes.

If on the other hand the areas are full, in the sense that the degree of environmental fluctuation dictates the minimum distance between species and the maximum tolerable overlap, then not only the tropical increase in R and decrease in U will affect its number of species, but so also will the greater value of $n\bar{\alpha}$ tolerated in the tropics. In other words the tropics-temperate gradient will be somewhat steeper, with closer species packing, if habitats in both areas are full; whereas if species are still accumulating, the gradient will be less steep with about equal packing in tropics and temperate. At present the data are inadequate to discriminate between these hypotheses.

Tropical and Temperate Islands

An island off the Maine coast may have most of the bird species it would have if it were embedded in the mainland; the Channel Islands off California have from 15% to 40% of the species they would have on the mainland; and the islands off Panama may have less than one-quarter of what they would have as part of the mainland. We can easily guess the reason: Temperate birds often migrate, and even those that do not usually move about a good deal. What is 20 more miles across the ocean to a bird that has already migrated from Colombia to Maine? A larger fraction of California birds are sedentary (Fig. 8-4) and would probably be somewhat more reluctant to colonize. Few Panama birds migrate anywhere and those that do (yellow-green vireo, *Vireo flavoviridis*; streaked flycatcher, *Myiodynastes maculatus*) are usually present on islands. But many Panama birds are unbelievably sedentary. For instance, small islands in Gatun Lake in the Canal Zone are often missing all antbirds in spite of the fact that these may be the commonest birds on the mainland just a few hundred yards away and in similar habitats. Some of these birds simply refuse to cross even small water gaps. J. M. Diamond (pers. comm.) says that the 10-m-wide water gap between New Guinea and Admosin Island has stopped about half of the New Guinea birds of appropriate habitat preference.

Diamond (1972) has compared actual immigration and extinction rates on a tropical island (Karkar, off New Guinea) with a temperate island (Santa Cruz, off Southern California). These islands are similar in area: Karkar, 142 square miles, and Santa Cruz, 96 square miles; they are roughly equal in distance from the mainland: Karkar, 10 miles from New Guinea, and Santa Cruz, 19 miles from the California coast. Diamond resurveyed Santa Cruz 51 years after a previous survey (see p. 79) and Karkar 55 years after a previous survey. During the interval between surveys, Santa Cruz lost 6 of 36 species (0.32%/year) and Karkar lost at least 5 of 43 (0.20%/year). Since Karkar is larger and would be expected to have a lower extinction rate, Diamond suggests islands of the same size in tropics and temperate would have roughly equal extinction rates.

Six species immigrated, unaided by man, onto Santa Cruz, and 11 new species appeared to be new on Karkar. Some Karkar immigrants were doubtless artifacts since the earlier study did not record birds only seen, so some large hawks that were doubtless formerly present appear as "immigrants." The 6 Santa Cruz immigrants came from a mainland fauna of about 93 species (land bird species living in lowlands and foothills of coastal southern California), and 32 of the 93 were already present on Santa Cruz. Hence 61 potential immigrant species provided 6 actual, known immigrant species in 51 years. This gives a temperate immigration rate of $\frac{6}{61 \times 51} = 0.19\%$ of the potential mainland species pool per year. Karkar was colonized, on the other hand, from a mainland pool of 228 species, 195 of which were not already present on the island. Therefore $\frac{11}{195 \times 55} = 0.10\%$ of the potential mainland species pool per year immigrated onto Karkar from New Guinea. Diamond uses a more sophisticated formula and gets very slightly different percentages. Thus while extinction rates for tropical and temperate islands seem to be about equal, the tropical immigration rate is only about half as great as the temperate rate. This "immigration rate" is not as viewed from the island, the rate we plot in later figures in this chapter, but rather the rate as viewed from the mainland. It gives the probability that a given mainland species will emigrate successfully in a given year. If there are many mainland species, the island will see a high total rate of immigration even if each mainland species is unlikely to go.

Patterns of Plant Morphology and Energetics

A visitor arriving in the tropics notices first, perhaps out of the plane window, the rather light metallic green of the foliage compared to temperate foliage. On landing he smells the various fungi that make the tropics smell like a moist greenhouse. On closer examination, the leaves of the forest trees are puzzlingly uniform to one used to using leaf shape to identify temperate trees. Very few leaves of trees of the forest interior have lobed leaves like our oaks, or stellate leaves like our sweet gums, or leaves with toothed margins like our elms. Instead, almost all the leaves seem to be elliptical with smooth margins and, perhaps, an acute tip. The bark too seems strikingly uniform. In fact, the experienced tropical forester will hack the tree with his machete and note the color and consistency of the sap that oozes out, and the color of the cambium layer. He finds these, plus the form of the buttresses around the base of the tree, to be the most useful identification points. But why are temperate leaves so irregular? Or why are tropical ones so regular? The question has never been properly answered, but we have some hints.

First, there are tropical leaves of irregular shape, but these are almost always in second growth forest understory or clearings. The giant palmate leaves of *Cecropia* are among the first learned by the visitor, but this is a tree of rapid growth confined to second growth forests and recently abandoned clearings.

Second, some temperate families like the oaks are lobed in some species and smooth in outline in others. The smooth oaks are almost without exception "evergreen," and the lobed ones are, without exception, "deciduous." Evergreen oaks in the United States keep their leaves for slightly more than 12 months compared to the 6 or 8 months that the lobed, deciduous leaves stay on. Furthermore the lobed oak leaves are usually considerably larger and more papery in consistency than the small leathery leaves of the evergreen oaks, leaves with smooth margin.

Third, the leaves on trees on a tropical mountain do not tend to be much more lobed than those in the lowland. Where this book is being written, in the forests of Vermont, there is not one species of tree with leaves of smooth outline. A tropical mountain with the same average annual temperature will have most of its tree leaves smooth margined.

Fourth, both the tropics and the temperate regions have trees with compound leaves whose leaflets are smooth. In the temperate zones these trees with compound leaves come into leaf later, and generally lose their leaves sooner, than the trees with entire leaves. Trees with compound leaves increase in number in arid areas but for present purposes we neither call these compound leaves "lobed" nor of smooth margin, but we leave them out of the comparison.

As a very tentative working hypothesis we might conjecture that lobed leaves are usually found on species that need seasons of rapid growth and production. Perhaps a lobed leaf is a natural consequence of very rapid leaf growth, and the papery thinness is incorporated because the leaf will be thrown away so soon. Conceivably also, in the tropics, where dry seasons are important, it is worth building a leaf with extra drought resistance and these are more leathery.

A related comparison of plant energetics in the tropics and the temperate has been made by Jordan (1971). He first points out that trees in tropical evergreen forests produce as much dry weight of leaf litter per year as they do of new wood. In the temperate, by contrast, forest trees produce between 1.5 and 6 times as much dry weight of new wood as of leaves. A tropical seasonal forest is more like the temperate in this respect. This result suggests again that in seasonal areas plants produce throwaway leaves that are as light as possible. This is borne out by the fact that wood production stays roughly constant (900–1000 gm/m^2/year) from tropics to temperate, whereas the fall of leaf litter is usually over 1000 gm/m^2/year in the tropics and reduces to 200–400 gm/m^2/year in the temperate and the tropical seasonal forest. Jordan also shows that the average energy content (in calories per gram of dry weight) is lower both in tropical leaves and entire tropical plants than in the temperate. Values range from 3000 to 4000 in the tropics and from 4500 to over 5000 in temperate and alpine areas. This pattern is puzzling since the fuel value of hardwoods is well known to be much higher than that of softwoods, and yet he feels values are higher for softwoods when the whole plant is considered. In any case tropical trees appear to store less energy in each gram than do temperate ones.

Modes of Selection: "*r* and *K* Selection"

We saw in the discussion of desert annual plants that there is no single strategy preferred by evolution. Rather, a variety of mutations enter the population and different mutant types are likely to prosper where the vagaries of the environment happen to coincide with the plants' adaptation.

It has sometimes been claimed that the one real strategy that guides evolution is avoidance of going extinct. It is of course a truism that, viewing evolution in retrospect, we see only the descendants of those whose lines did not go extinct. However, if we examine any present-day organism there is no evidence that it is adapted to minimize its chances of future extinction. What one sees is a set of adaptations, made in the evolutionary past, that have proved workable up to the present moment. A more mechanistic view helps us here. Mutations are continually appearing in the population; and unless the population is continually exposed to the risks of imminent extinction, there is no machinery that will save just those mutations that will prevent future extinction. In other words, a study of the machinery of the incorporation of new mutants will save us from useless platitudes about the mutations a population would preserve if it only were foresightful enough. Finally, we observe that to minimize the chances of a local extinction of part of a population is not the same as to minimize the chances of a complete extinction of the total population. A population with many independent subpopulations, each with a high extinction probability, is relatively safe from total extinction if the recolonization rates of areas occupied by populations gone extinct are high; and if recolonization rates are very low, the total population is in danger of extinction even if its subpopulations have low extinction rates.

Part of classical population genetics is just an elaborate restatement of the fact that if X_1 and X_2 are the populations of two alleles and if $\frac{1}{X_1}\frac{dX_1}{dt} > \frac{1}{X_2}\frac{dX_2}{dt}$, then X_1 is gaining on X_2. If, further, $\frac{1}{X_1}\frac{dX_1}{dt}$ stays greater than $\frac{1}{X_2}\frac{dX_2}{dt}$, X_1 will eventually form virtually all of the combined population. These results are routine because the interesting biology is hidden in the expression $\frac{1}{X}\frac{dX}{dt}$. The population geneticist is content to say, "Let fitness $= \frac{1}{X_1}\frac{dX_1}{dt}$" and he then proceeds in a purely mathematical program. The ecologist can do slightly better because he further believes $\frac{1}{X_1}\frac{dX_1}{dt}$ begins at a high level r and, as X_1 grows to some

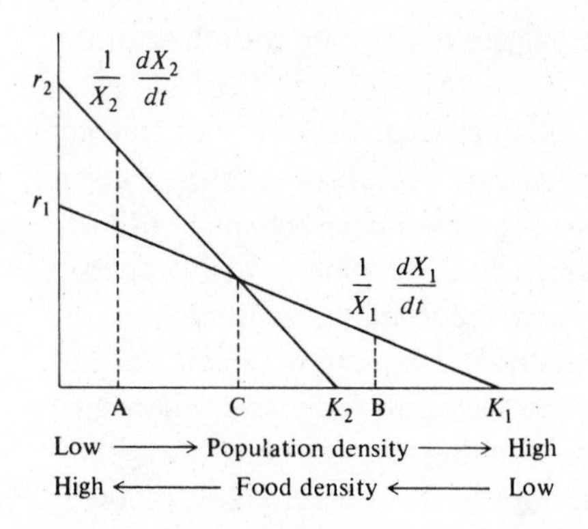

Fig. 8-15

The rates of increase of two alleles X_1 and X_2 are portrayed as functions of population density. X_1 increases faster than X_2 when density exceeds C, but X_2 increases faster when the density is less than C. Since high food density is often correlated with low population density, this coordinate is also shown. See the text for details.

level K, $\frac{1}{X_1}\frac{dX_1}{dt}$ declines to zero. In many cases $\frac{1}{X_1}\frac{dX_1}{dt} > \frac{1}{X_2}\frac{dX_2}{dt}$ not only at low population densities but also at high densities. Then the ecologist has little new to contribute. But when $\frac{1}{X_1}\frac{dX_1}{dt}$ and $\frac{1}{X_2}\frac{dX_2}{dt}$ cross as in Fig. 8-15, then there is something new. (For cases where $\frac{1}{X_1}\frac{dX_1}{dt}$ and $\frac{1}{X_2}\frac{dX_2}{dt}$ are not uniquely determined by total density, see the graphs in MacArthur and Wilson (1967).) We contrast populations of the two alleles in two different environments. In the first environment the food supply and population density are those corresponding to point A in the figure. Above this point $\frac{1}{X_2}\frac{dX_2}{dt} > \frac{1}{X_1}\frac{dX_1}{dt}$ so X_2 is winning. This is often called "*r* selection" because the r_1 and r_2 of the competition equations are the left intercepts of the $\frac{1}{X}\frac{dX}{dt}$ lines; the allele with the greater r is winning. In an environment with food and density corresponding to point B in the graph $\frac{1}{X_1}\frac{dX_1}{dt} > \frac{1}{X_2}\frac{dX_2}{dt}$ so X_1 is winning. This has been called "*K* selection" because the right intercepts are K's of the competition equations and here the allele with the larger K is winning; r selection switches to K selection where the lines cross. For a different trait the switchover will usually occur at a different density. If the lines do not cross, as must frequently happen, r and K selection coincide. Roughgarden (1971) has given a more detailed account of these ideas in a slightly different way.

We can tentatively relate these concepts to our tropical-temperate comparisons as follows. In the relatively nonseasonal tropics, populations may stay large and keep their food supply low. These tropical environments are then like point B in the figure, and K selection prevails. In very seasonal temperate environments population growth may be saw-toothed, with exponential growth followed by catastrophic decline caused by storms or migration. If the points of these sawteeth are below the density of switchover from r to K selection, these temperate regions will have r selection. Of course a population may easily spend half its time undergoing r selection and the other half with K selection; all mixtures are possible. The species that occupy very temporary habitats, such as the plants of early stages of succession, or the animals that must perennially recolonize small islands of a large random extinction rate, will tend to be r-selected.

It is of interest to think of alternative traits whose $\frac{1}{X}\frac{dX}{dt}$ lines might cross as in the figure; such traits only will have opposing r and K selection. We begin with a hypothetical example which, however, clearly illustrates the kind of situation to look for. Suppose a mutation arose in a grazing mammal which pointed the incisor teeth outward. We suppose this mutation lets its bearers graze the grass closer because their teeth meet closer to the ground. However, the teeth are now poorer at shearing grass. Thus when grazers are scarce and grass is long (r selection) we would expect the normal tooth type to harvest more food and produce more offspring; when grazers are common and grass has been cropped very short (K selection), the mutant type that can graze closer will win; in fact then the mutant type may graze so close that the normals cannot feed at all.

Odum and Pinkerton (1955) gave many examples that illustrate why large productivity and large efficiency are incompatible. The essential reason is that production is measured per unit time so that large productivity means rapid even if wasteful conversion of food into body material. Efficiency is a ratio of power output to power input; since it is a ratio, it is not measured per unit time and high speed does not increase efficiency. In fact, very high speed is likely to produce turbulence and decrease efficiency. To this extent selection in the tropics may be conjectured to favor efficiency, and selection in the temperate zones to favor productivity. Roughgarden (1971), however, shows that *K* selection favors not efficiency per se but insensitivity of production to population density.

We consider another example, genotypes for large and small clutch sizes in birds. The large clutch genotype will, if all young survive, produce faster population growth. But when food is scarce there will be fewer survivors from a large clutch in which some will die and have food wasted on them than from a small clutch all of whose members are well fed. Hence tropical *K* selection would favor small clutches whereas *r* selection would favor large ones. This does not explain the large clutches of the hole nesters, however (pp. 220–221). The same should hold true in plants: those in temporary environments (*r* selection) should devote a larger proportion of their energy to reproduction than should *K*-selected plants in more stable environments. M. Gadgil and O. Solbrig (pers. comm.) have shown this for dandelions, in which plants of unstable habitats put more energy into flowers than those from stable habitats.

The subdivision of all natural selection into *r* selection and *K* selection is convenient because it is a fairly natural subdivision, but it is by no means the only possible one. We could speak of youth vs. old-age selection; or predator escape selection vs. selection for feeding ability, and so on. It is too early to assess the role that these other subdivisions will play in our understanding of evolution, but their investigation will probably be rewarding.

Appendix 1 *Theory of Patchy Distributions Under Competition*

It is the purpose of this section to extend the results of the appendix to Chapter 2 to include a whole community of competitors. In Chapter 2 (pp. 38–39) we showed that, if there are two resources and two consumers, then at equilibrium the species abundances X_1, X_2 satisfy

$$0 = a_{11}w_1K_1 + a_{12}w_2K_2 - T_1 - \left(a_{11}^2 w_1 \frac{K_1}{r_1} + a_{12}^2 w_2 \frac{K_2}{r_2}\right)X_1 - \left(a_{11}a_{21}w_1 \frac{K_1}{r_1} + a_{12}a_{22}w_2 \frac{K_2}{r_2}\right)X_2$$

$$0 = a_{21}w_1K_1 + a_{22}w_2K_2 - T_2 - \left(a_{11}a_{21}w_1 \frac{K_1}{r_1} + a_{12}a_{22}w_2 \frac{K_2}{r_2}\right)X_1 - \left(a_{21}^2 w_1 \frac{K_1}{r_1} + a_{22}^2 w_2 \frac{K_2}{r_2}\right)X_2 \qquad (1)$$

Here we carry over the assumptions used in the derivation of Eqs. (1). In particular, the resources must renew in a specified fashion. Under two further assumptions we can cast this equation into a new and useful form. Suppose $T_1 = T_2 = T$, the species have equal metabolic needs, and $a_{11} + a_{12} = a_{21} + a_{22} = a$, so that the species' total harvesting abilities are equal. Then we write an expression whose virtues will be apparent presently. Let Q be given by

$$Q = \frac{w_1K_1}{r_1}\left(r_1 - \frac{Tr_1}{K_1a} - a_{11}X_1 - a_{21}X_2\right)^2 + \frac{w_2K_2}{r_2}\left(r_2 - \frac{Tr_2}{K_2a} - a_{12}X_1 - a_{22}X_2\right)^2 \qquad (2)$$

In order to find the values of X_1 and X_2 that minimize Q we would equate $\frac{\partial Q}{\partial X_1}$ and $\frac{\partial Q}{\partial X_2}$ to zero: $0 = \frac{\partial Q}{\partial X_1} = -2a_{11}\frac{w_1K_1}{r_1}\left(r_1 - \frac{Tr_1}{K_1a} - a_{11}X_1 - a_{21}X_2\right) - 2a_{12}\frac{w_2K_2}{r_2}\left(r_2 - \frac{Tr_2}{K_2a} - a_{12}X_1 - a_{22}X_2\right)$. Dividing by

-2 and collecting constant terms and coefficients of X_1 and X_2, this becomes $0 = a_{11}w_1K_1 + a_{12}w_2K_2 - \left(\frac{a_{11}+a_{12}}{a}\right)T - \left(a_{11}{}^2w_1\frac{K_1}{r_1} + a_{12}{}^2w_2\frac{K_2}{r_2}\right)X_1 - \left(a_{11}a_{21}w_1\frac{K_1}{r_1} - a_{12}a_{22}w_2\frac{K_2}{r_2}\right)X_2$, which is the first of Eqs. (1)! Similarly, setting $0 = \frac{\partial Q}{\partial X_2}$ gives us the second of Eqs. (1). In other words, the equilibrium values of X_1 and X_2 are those numbers X_1 and X_2 that minimize Q. We can conclude that competition results in species abundances that minimize Q. That it really is a minimum and not some other stationary point of Q is shown formally by MacArthur (1970); it should be obvious biologically when we picture Q.

Q can be easily extended to the case of n species competing for m resources. Then the equilibrium abundances $X_1, X_2, X_3, \ldots, X_n$ of the species are those that minimize $Q = \sum_{j=1}^{m} \frac{w_jK_j}{rj}\left(r_j - \frac{r_jT}{K_ja} - \sum_{i=1}^{n} a_{ij}X_i\right)^2$. The term in parentheses is the difference between what might be called the "useful production of resource j," $r_j - \frac{r_jT}{K_ja} = r_j\left(1 - \frac{T}{K_ja}\right)$, and the "utilization of resource j," $a_{1j}X_1 + a_{2j}X_2 + \cdots + a_{nj}X_n$. Thus we can rephrase the conclusion by saying that the equilibrium abundances $X_1, X_2, \ldots, X_n$ are those which minimize the weighted $\left(\text{by } \frac{WK}{r} \text{ terms}\right)$ squared deviation between "useful production" and "utilization of resources." In particular if there are values $X_1, \ldots, X_n$ such that

$$a_{1j}X_1 + a_{2j}X_2 + \cdots + a_{nj}X_n = r_j\left(1 - \frac{T}{K_ja}\right) \tag{3}$$

for all j values, then the term in parentheses is zero for all j, and Q is zero, which is its minimum possible value. Species of abundances that satisfy Eq. (3) then are a noninvadable community, for the addition of an independent term $a_{n+1,j}X_{n+1}$ would increase Q to a higher value. Why does this mean the original community cannot be invaded by species $n+1$? The equilibrium of all $n+1$ species is the least value Q can take; we know that with all $n+1$ species $Q > 0$ while with only the first n, Q can equal zero. Hence the $n + 1$ species will equilibrate with the $n + 1$st missing.

We now apply this scheme to understanding patchy distributions and relative abundance. As we have often done before, we plot the resources, j, as a continuum and the species utilizations $a_{1j}, a_{2j}, \ldots, a_{nj}$ as n distinct curves above this coordinate. For convenience we make them triangular so that it will be easy to draw conclusions. Finally, we superimpose the curve of "useful production" $r_j\left(1 - \frac{T}{K_j a}\right)$ as in Fig. 8-16. Our results now say that the abundances $X_1, \ldots, X_n$ are those multiples of the $a_{1j}, \ldots, a_{nj}$ curves that in

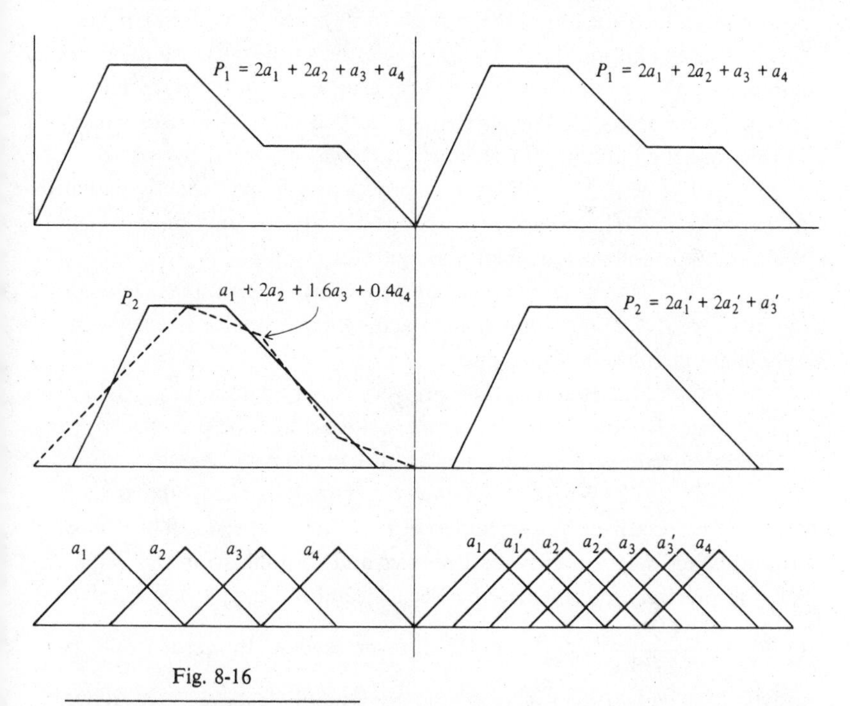

Fig. 8-16

Two regions of resource production, P_1 and P_2, are plotted as functions of resource. a_1, a_2, etc., are the utilizations of the consumer species 1, 2, etc. On the left, competition is low, $\alpha = 0.25$, and the species are separated by 2.44 standard deviations. On the right, the species are more closely packed, so that $\alpha = 0.7$, and the species are separated by 1.22 standard deviations. See the text for the interpretation.

combination give the best match to the production, and that if some positive multiples of the a_{ij} curves exactly match the useful production, that community cannot be invaded. For example, the upper production curves (per hectare) in the figure are exactly matched by $2a_1 + 2a_2 + a_3 + a_4$, so species 1, 2, 3, and 4 will be present in abundances 2, 2, 1, and 1 per hectare. The lower production curve, P_2, cannot be perfectly matched by these four species but Q is minimum when $X_1 = 1$, $X_2 = 2$, $X_3 = 1.6$, and $X_4 = 0.4$, approximately, as shown in the figure. Notice that resource utilization can exceed "useful production," $r_j - \frac{r_j T}{K_j a}$, so long as the utilization does not exceed the greater value, r_j. Any resources for which minimizing Q leads to utilization exceeding r_j will go extinct. In this case we must form a new Q without the extinct resource and minimize it. If a species gets a negative abundance on minimizing Q, we eliminate it and form a new Q. On the right side of the figure the only difference is that species are packed twice as closely. In this case the habitat with production P_1 was occupied by species 1, 2, 3, and 4, while the habitat with production P_2 was exactly matched by $2a_1' + 2a_2' + a_3'$ and species 1′, 2′, and 3′ are the only ones present. We draw four conclusions from the figure.

1. Where species are not closely packed (e.g., left side of figure where $d = 2.44\sigma$), similar habitats will have similar species with only slight changes in abundance.

2. Where species are very closely packed (e.g., right side of figure where $d = 1.22\sigma$), similar habitats may be occupied by very different collections of species; that is, distributions are patchy.

3. Where species are very closely packed, similar habitats may be occupied by very different numbers of species. This follows from the fact that P_2 holds three species and P_1 holds four and yet a mixture precisely intermediate between P_1 and P_2—i.e., a new P_3 given by $P_3 = \frac{P_1 + P_2}{2}$—would contain all seven species. In fact it would be exactly matched by $a_1 + a_1' + a_2 + a_2' + 0.5a_3 + 0.5a_3' + 0.5a_4$. Hence a small change in habitat could cause a large change (from three to seven) in numbers of species, although in most cases there would be between five and seven species. The left-hand community would contain four species in all these habitats.

4. When the environment is seasonal, with P_1 the production in one season and P_2 in another, then the left-hand community would persist with all species present, while the right-hand community would have species go extinct. If it reached P_1 first, then only species 1, 2, 3, and 4 would remain; if it reached P_2 first, then only 1′, 2′, and 3′ would remain. In either case the fluctuating environment has made the packing of the right-hand community intolerably tight. Hence a fluctuating environment sets a limit to the closeness of packing; the greater the fluctuations, the less close the packing can be.

Appendix 2
Species Packing and Crystal Packing

Most people have noticed that footprints in wet sand become dry. By stepping on the sand one rearranges the sand grains that were previously packed with minimal spacing between the grains. The new arrangement is not so efficiently packed, and hence the surface of the sand is raised above that of its surroundings. More generally, an assemblage of crystals can normally be given an arrangement that minimizes the interstices between crystals. Any alteration of this arrangement will then reduce the closeness of the packing.

We have just seen that certain competitive assemblages of species reach equilibrium when their abundances are such that resource utilization best matches the useful production. As various new species attempt to invade, only those succeed that can improve the matching, and a typical mainland community can be viewed as one that has tried all locally available species and has kept those that will improve the matching. It seems quite likely, although difficult to prove, that as a result of this sifting of available species not only will the number of individuals of all species combined often be maximized, but also that the species will often, in the end, be fairly uniformly spaced with the maximum possible number of coexisting species present.

When this happens—when competitive packing like crystal packing has led to maximum density of individuals and maximum number of species—then we can expect certain communities to hold fewer individuals and species. Thus, for instance, the ten species that are actually on some island are probably not packed as well as some ten mainland species might be. For this reason, islands may be less efficiently occupied than is possible. Again, newly recolonized islands such as Wilson and Simberloff's E1 (p. 82) may have less-well-packed species than the island had originally, which would explain why the recolonization seems to equilibrate with fewer species. (I am indebted to Gordon Lark for this suggestion. Lark also suggests that large islands, especially land bridge islands that had a wide variety of mainland species, might achieve better packing than a mainland in which spillover of species from adjacent habitats interferes with the achievement of close packing.) Conceivably the high tropical species diversities reflect, in part, a closer competitive packing. Levins (1968) and Vandermeer (1970) describe species diversity with mathematics especially suitable for understanding the crystal-packing effects.

Such crystal-packing effects are second order effects. That is to say, the effect of shifting from close packing to some other kind is to change numbers of individuals and species by small percentages. The main, first order causes of species numbers are those we have described in Chapter 7, Eqs. (1) and (2).

The Role of History

9

Unravelling the history of a phenomenon has always appealed to some people and describing the machinery of the phenomenon to others. In both processes generalizations can be made and tested against new information so both are scientific, but the same person seldom excels at both. The ecologist and the physical scientist tend to be machinery oriented, whereas the paleontologist and most biogeographers tend to be history oriented. They tend to notice different things about nature. The historian often pays special attention to *differences* between phenomena, because they may shed light on the history. He may ask why the New World tropics have toucans and hummingbirds and parts of the Old World have hornbills and sunbirds. The machinery person may instead wonder why hummingbirds and sunbirds, despite their different ancestries, are so similar. He tends to see *similarities* among phenomena, because they reveal regularities. However, many generalizations about machinery have been proposed by biogeographers with an historical outlook, and here we review these and some other generalizations about historical biogeography. We also review where history leaves an indelible mark even upon the equilibria so dear to the ecologist. Of course, the time scale of history can be measured in decades or in billions of years. We begin with the decades.

Patterns in History

In a remarkable book, Hastings and Turner (1965) have described the history of the changing vegetation in southern Arizona. Their basic method was to locate old photographs of the Arizona landscape, usually taken around 1900. They then carefully relocated the places where the pictures were taken and took new ones. They exhibit both pictures of each location, commenting on what kinds of plants there were and which had died or appeared between the dates of the two pictures. The overall effect is overwhelming. We see hillside after hillside in which oak savannah is slowly dying back and mesquite (*Prosopis*) is invading. Mesquite and some associated plants are also taking over what had been grassland, and the giant saguaro cactus (*Cereus giganteus*) of the lower deserts is dying out of some locations because of lack of reproduction. These changes began at about the time the intensity of cattle grazing reached its peak in these very habitats, although there may have been independent climate variations helping to promote the vegetation change. To show how intricate the chain connecting cattle grazing to lack of baby saguaros can be, we retell the story of the difficulties of the saguaro related by Niering, Whittaker, and Lowe (1963). First, some baby saguaros are killed by cattle either eating them or stepping on them. However, many young saguaros do get established under thorny trees that protect them from direct cattle damage, but even these are not found where there was heavy grazing. Second, desert rodents, especially woodrats of the genus *Neotoma*, eat cactus (unlike the kangaroo rats and pocket mice, they need moisture and tend to be associated with cactus, mostly smaller cactus like prickly pear and cholla, both of the genus *Opuntia*); "protection" of cattle against predators means killing coyotes and a consequent increase in the woodrats and decrease of young saguaros. Third, cattle grazing changes the plant composition of the desert by selectively reducing the seed-producing annual plants favored by other rodents and favoring the *Opuntia* cactuses, thus providing another reason why woodrats should be common and saguaros endangered. In any case, Niering, Whittaker, and Lowe show that, empirically, rodents and especially woodrats are commoner in the grazed areas, presumably causing much of the trouble. It is a grim reminder of man's ignorance that the huge "Saguaro National Monument" set aside east of Tucson to save and exhibit these cactuses for posterity is the place where the dying back is fastest; grazing was allowed in the monument until 1958 and it may be too late to save the saguaros. To the government's credit, it may be added that a new, ungrazed Saguaro National Monument west of Tucson has been set aside. Here not only is the desert even richer and more interesting, but the saguaros are reproducing very well.

The next time scale we deal with is measured in few thousands of years. This time scale is important for two reasons. First, between 12,000 and 20,000 years ago the most recent glaciation covered northern parts of the Northern Hemisphere, lowered the water level of the world's oceans by about 100 meters, and doubtless changed the climates even of tropical lands. Second, the length of time it normally takes for a species to split and diverge sufficiently to be regarded as two species is a small, uncertain number of thousands of years. In other words, there have been dramatic changes in climate and vegetation over tens of thousands of years, and that is a sufficient length of time for events as important as the formation of new species.

Plants cover a much greater part of the earth than do animals; hence if plants fossilized well their remains would leave an incomparable record of the past. One part of plants only, their pollen grains, seem nearly indestructible—at least in acid soils—and here there is an opportunity to understand the plant geography of the past. For more than 40 years people have been taking cylindrical cores of peat with embedded pollen grains from acid bogs and lake bottoms. With some practice and patience they have learned to identify the plant that produced the pollen, sometimes even to species, sometimes to genus, and almost always at least to family. Hence, the pollen analyst can infer that his pollen sample was from oaks, pines, or grass and sedges, and so on. Unfortunately, different trees produce different amounts of pollen and disperse it to different distances. Pines, for instance, produce far more than most other trees. Consequently, it is not straightforward to infer the quantitative composition of the flora from the percentages of different pollen grains. Davis (1969) has solved this problem by putting out trays to catch pollen in different kinds of forest. When she finds a present-day forest whose pollen percentages closely correspond to those in the pollen from a section of a lake bottom, she infers a similarity of the flora that produced the pollen in the lake bottom to the flora in the present forest. Using radiocarbon dating of the peat or wood remains surrounding the pollen, one can also infer the age of this pollen assemblage. By a combination of these techniques Davis has strengthened the earlier conclusions of other pollen analysts that there was a substantial strip of tundra along the edges of the glaciers 12,000 years ago when they were just beginning their retreat (Fig. 9-1).

In the tropics pollen studies of this kind are in a more primitive state, partly because of the difficulty in assembling a museum of pollen grains for identification, but results of great interest have already been achieved. For instance, van der Hammen and Bigarella (both cited by Vanzolini and Williams, 1970) made studies in Colombia and coastal

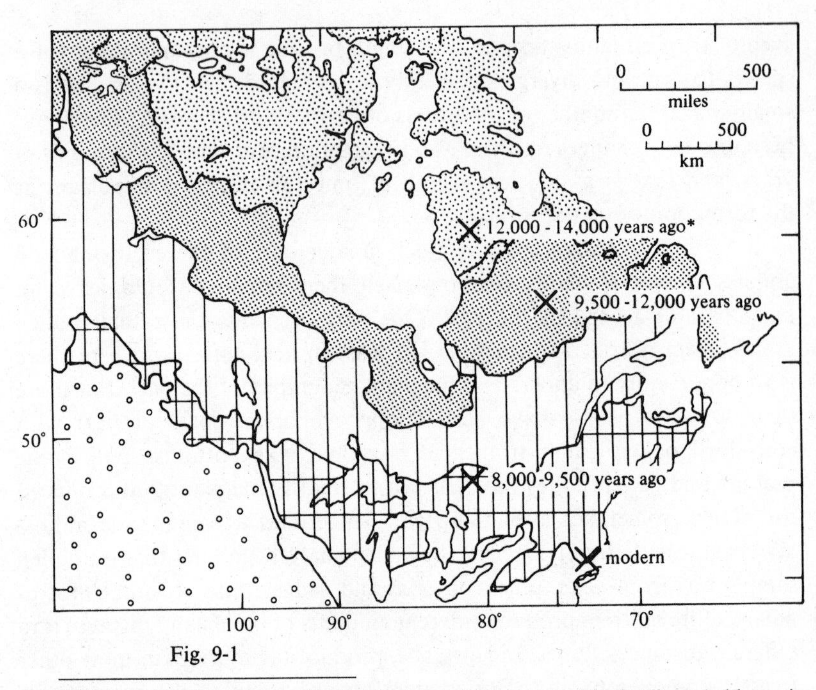

Fig. 9-1

Localities in Canada where surface pollen assemblages resemble fossil assemblages in southern New England. The sites are indicated by X's. The age of the analogous fossil material is indicated for each locality. (Location in the tundra indicated by the asterisk (*) is extrapolated from the resemblance of fossil material to assemblages from northernmost Quebec. Surface samples are not yet available from the precise locality shown.) (From Davis, 1969.)

southeastern Brazil, respectively, and found that grasses showed a great increase (relative to the present) between about 3500 years ago and 2000 years ago. Vanzolini and Williams also site geological evidence of dry periods in parts of the Amazon forests where soils are found that seem only to be produced in a dry climate. Finally, Vanzolini and Williams infer "core" areas for the evolution of the *Anolis chrysolepis* species group corresponding to the areas that escaped the droughts. These core areas are regions where the value of characters of the species complex are relatively uniform and yet are different from the characters of the species in the other core areas. These inferred core areas from which they postulate evolutionary radiation correspond quite closely to the areas that presumably remained wet even during the dry periods of recent geological history. The significance of this reasoning is that it demonstrates how species could have formed even in what appear to be the uniform, stable tropics. The ecologist never doubted that there have been many opportunities for speciation, but it is reassuring to know just how it probably happened. Haffer (1969) independently inferred very similar core area or "refugia" for the evolution of bird superspecies by considering where the semispecies now occur.

Such refugia have been proposed in excessive detail for North America. There is no doubt that during glacial times the vegetation map of North America looked different. For instance, the southeast, Texas, and the southwest are often viewed as warm refugia into which temperate species were forced by the advancing glaciers. In these refuges, the story goes, the isolation that enabled North American species to speciate took place. So far so good. But by invoking several refugia over several glaciations one could account in too many different ways for the existing species. Any selection among these alternatives is purely arbitrary. Furthermore, interglacial periods were longer than glacial periods, and the present vegetation map offers many opportunities for speciation. Why, then, try to explain everything in terms of glacial refuges—that is, vegetation islands during glacial periods? Why not explain part of what has taken place in terms of vegetation islands in the interglacial periods? Present distributions of related species simply do not contain enough information to discriminate between the alternatives.

When we measure time on a still longer scale, new phenomena appear. Over millions of years continents drift apart, and whole faunas and floras appear to take vast migrations. Sometimes whole faunas invade other ones, causing gigantic experiments in competition, and sometimes there are vast extinctions. The events of this time scale used to be considered the whole science of biogeography. A very good sample of this kind of work is provided by Darlington's book *The Biogeography of the Southern End of the World* (1965), the aim of which is to unravel the history of faunal and floral similarities and differences between southern South America and southern New Zealand and Tasmania. To proceed with his task, Darlington distils the following operating principles (partly anticipated by Darwin):

1. There are observed to be more species in continents of larger area and milder or "more favorable"—i.e., tropical—climate.

2. Colonists from these species-rich areas are more likely to succeed than those from smaller continents in less favorable climates. Not only are the colonists more successful but the net dispersal is greater from large continents in favorable climates.

The first of Darlington's working principles is an empirical fact that cannot easily be disputed as long as "favorable" is interpreted as applying to the tropics. The second is a principle of great interest. It appears never to have been confirmed in a statistically satisfactory way, for large continents have more species and individuals that can attempt a colonization, and a net drift from large to small does not imply, in itself, any competitive superiority of the colonists from the large continent. However, the principle is so plausible that it is very likely to be true. The adage "Practice makes perfect" should apply to competition as well as to other activities, and the emigrants from species-rich continents in tropical climates have had much practice in competing. They certainly should be good at invading a new community of competitors. This is the presumed reason why more North American mammals invaded and succeeded in South America than the reverse, when the Central American isthmus was closed in the Pliocene, and why more Eurasian species invaded and succeeded in North America than the reverse when faunas were exchanged across the Bering Sea.

Notice, however, that a species that is good at invading a continent with an existing community of competitors need not be good at colonizing a remote island. We have examined the properties of good island colonizers, and ability to compete played a role only when the island held a fair number of species. In the early stages other properties were more important, and for these islands there seems to be little a priori reason to expect Darlington's principle to hold. In fact, Williams (1969) has documented the invasion of an *Anolis carolinensis* ancestor from South America into the West Indies, where it thrived and speciated, and thence into southeastern United States, where it met few competitors. Thus, a species living on an island successfully invaded a large continent, but a part of a large continent with few competitors.

Darlington raises the following question: Why are virtually no warm-blooded terrestrial vertebrates shared by and confined to southern South America, New Zealand, and Australia, whereas many plants and invertebrates are? Why, in fact—and this is a partial answer to the first question—are there so few South American vertebrates confined to cold temperate regions? For the most part, southern South America has vertebrates that are also in the tropical part of that continent and appear to have adapted to progressively colder climates. Darlington concludes that warm-blooded vertebrates are better at moving into new regions than are insects or plants and that consequently fewer relict vertebrate species, left from an earlier era, remain. With a slight change from Darlington's emphasis we can recast this explanation in terms of island theory. The small southern ends of southern continents offer some difficulties (cold) to the immigrant and are too small for much effective vertebrate speciation. As a result, the rate of addition of new species of vertebrate is small. The rate of extinction of vertebrate species would be large since vertebrates have low densities and so would be relatively rare in such a small area. Hence the vertebrate equilibrium will be for few species with a high turnover rate (Fig. 9-2), with no relicts or endemism. Invertebrates should maintain vastly denser populations and therefore have lower extinction rates.

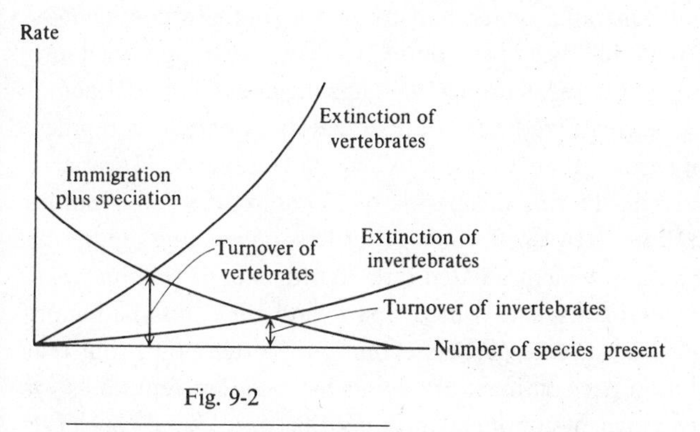

Fig. 9-2

The balance of extinction against immigration plus speciation in temperate South American vertebrates and invertebrates. Since vertebrates are rarer, they presumably have higher extinction rates. Hence the turnover rate, measured in extinctions per unit time, is greater for vertebrates, which are therefore less likely to remain as relicts. Furthermore, speciation is likely to form a larger component for small invertebrates than for the large mobile vertebrates that cannot find room in southern South America for much geographic isolation. As a result, more invertebrates are likely to be endemic.

Also, being much smaller, they can probably occasionally speciate in the small bits of continents involved. On the other hand, it is most likely more difficult for an invertebrate adapted to the tropics to invade progressively the cold temperate, so invertebrate immigration rates may be less. In sum, the addition of invertebrate species may be about equal to that of vertebrates, but a higher proportion will have speciated on the spot and be candidates for endemism. Figure 9-2 shows that at equilibrium there would be a vastly higher turnover rate of vertebrates with consequent negligible chance for relict endemics.

In considering the earth's geography in those remote times, we come upon the vast land movements known as continental drift. Geologists currently view the earth's crust as composed of a few giant floating "plates" drifting on the currents or eddies of a liquid mantle. There are source areas where new plate material is formed and from which plates diverge, and there are sink areas where plates meet and plunge downwards, or one rides over another. Earthquakes and volcanoes are concentrated along plate boundaries. When land continents are situated on plates diverging from each other, the continents too drift apart. Thus North America and Eurasia drifted apart, and so did South America, Africa, India, and Australia, which had formed a southern continent of Gondwanaland. What had been two continents drifted and regrouped into at least six major land masses. Kurten (1959) has suggested that continental drift accounts for the greater radiation of mammals than, earlier, of reptiles. During the age of reptiles, there were but two major continents, "Laurasia" and "Gondwanaland." During this period of 75 million years, the reptiles only subdivided into between 7 and 13 orders (depending upon the taxonomist). Mammals, which radiated into 30 orders in 65 million years, did so after the continents had drifted apart, and they thus had more opportunities for isolation and parallel evolution. Kurten points out that the rates of evolution per continent are about the same for reptiles as for mammals. The two ancient continents produced between 7 and 13—let us say 10—orders of reptiles in 75 million years, which comes to 5 orders per continent. In a roughly equal time the mammals on six continents produced 30 orders; again 5 per continent. Of course, this result may be an accidental agreement and more dependent on what mammalogists and herpetologists consider to be an "order" than on continental drift.

Alternate Stable Equilibria

History even leaves its mark on equilibria, although how long its influence will be felt is unknown. We have already seen (p. 91) how very hard it is for a second species to colonize an island containing a reasonably close competitor. In this sense whichever species arrives first is practically permanent, and the later arrival is virtually certain to remain missing. But this is not really stable. Given enough time, early species A will go extinct from some islands and B will certainly successfully invade some of them. By this time random processes will have erased most of the history.

There is another way in which history can leave a fairly permanent mark. To see this, we must return now to the graphical presentation of competition that appears in the appendix to Chapter 2 (pp. 46–56). Further, in what follows in the balance of the chapter, we must draw freely both on the theoretical results described in the Chapter 2 appendix and on the extension to this theory derived in Appendix 1 to Chapter 8.

Here we have two resources, probably quite similar, and three consumer species harvesting them. Two species, X_1 and X_2, are far better at harvesting one than the other resource, and in this context are specialists. The third consumer species, X_3, is not only equally good at using both resources, but in fact very good at utilizing both, so that its isocline lies inside the intersection of the first two (see Fig. 9-3). Then the two resource levels, A and B, are alternate stable equilibria, each resistant to invasion by the remaining species. If X_2 and X_3 are present, they maintain the resources at level A and the X_1 isocline lies outside A, so X_1 cannot invade; similarly, if X_1 and X_3 are present at equilibrium, they reduce the resources to level B and X_2 cannot invade. It looks as if history leaves an indelible trace on this kind of situation, but in fact this situation is usually vulnerable to another fate. To see this, we note that if the X_3 isocline passed outside of point C, there would not be alternate stable equilibria. Then the only stable equilibrium point would be C; either of the other intersections would be vulnerable to invasion by the third species and would end up at point C, with only consumers X_1 and X_2 present. Hence, the condition for alternate stable equilibria is that the generalist isocline lie inside the intersection, C, of the specialist isoclines. A glance at Fig.

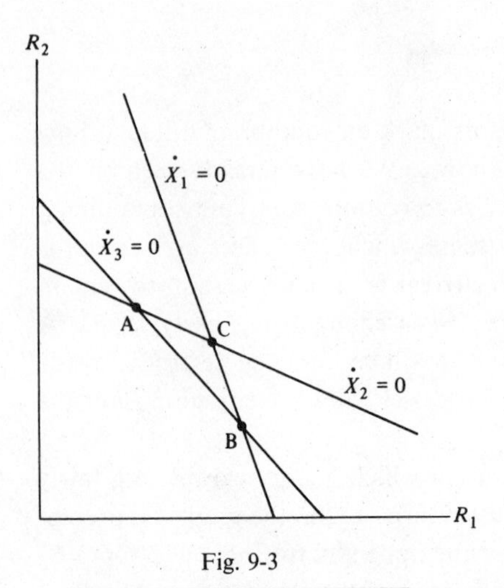

Fig. 9-3

Plotted as in Chapter 2, the isoclines $\frac{1}{X_1}\frac{dX_1}{dt}(=\dot{X}_1)=0$, $\frac{1}{X_2}\frac{dX_2}{dt}(=\dot{X}_2)=0$, and $\frac{1}{X_3}\frac{dX_3}{dt}(=\dot{X}_3)=0$. Since $\dot{X}_3=0$ lies inside the point C, X_3 can invade a community consisting of X_1 and X_2 and either eliminate X_2 and come to equilibrium with X_1 at point B or eliminate X_1 and come to equilibrium with X_2 at point A. Points A and B are alternate stable equilibria, each resistant to invasion by the remaining species.

2-13 shows that this condition is naturally met only when the resources are so similar that the consumers would have an evolutionary convergence. Thus, although we have exhibited the possibility of alternate equilibria resistant to ecological alterations, these equilibria are vulnerable to evolutionary alterations.

We can, however, construct interesting whole communities each resistant to invasion by species from the other. These appear stable both in an ecological and an evolutionary sense. To do this we use the format and assumptions of Appendix 1 to Chapter 8 and of Fig. 8-16, where we found that those species would be present and with such abundance that their total utilization most closely matched the "useful production" along the whole resource spectrum. In Fig. 9-4 we show the simplest case of these alternate communities. The two areas have just slightly different useful production curves, P_j, but each has its production curve exactly matched by a sum of the utilizations of the separate species

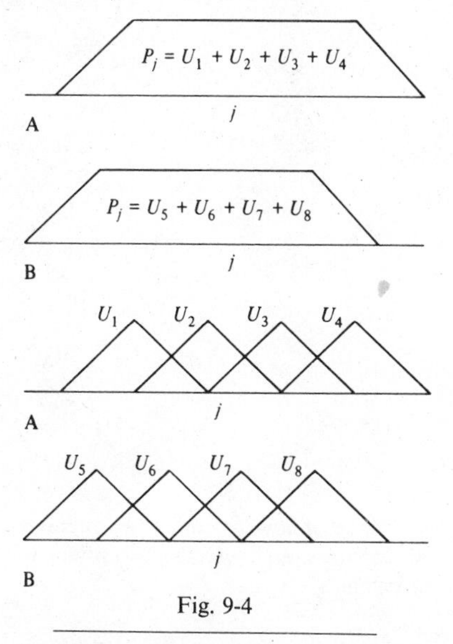

Fig. 9-4

Production curves P_j for regions A and B, and utilization curves U_i for species 1 to 4 of region A and species 5 to 8 of region B. The perfect matching of P_j by combinations of the four curves shows that each region is uninvadable by species of the other region. See the text for details.

and so each is uninvadable. However, the production curves overlap enough so that species 2 and 7 would each be able to live equally well in the other environment if there were no competitors. Their resources are equally present in both environments and yet neither can invade the other's environment. We can even go further and exhibit, slightly artificially, two alternate communities each occupying the same environment and each resistant to invasion by species from the other (see Fig. 9-5). This sort of situation is likely where two independently evolved faunas meet. In nature, it would not be quite as clear cut as in the figure because neither community would have its productions exactly matched by its utilizations and so each would be slightly, but only slightly, vulnerable to new invasions. Hence, where these two faunas meet we would expect each to maintain its integrity quite well, up to a very narrow zone of chaos where one is replaced by the other. This is diffuse competition in its purest sense. Let us consider a real case. A bird watcher in England or continental Eurasia finds warblers and flycatchers of many species breeding in the

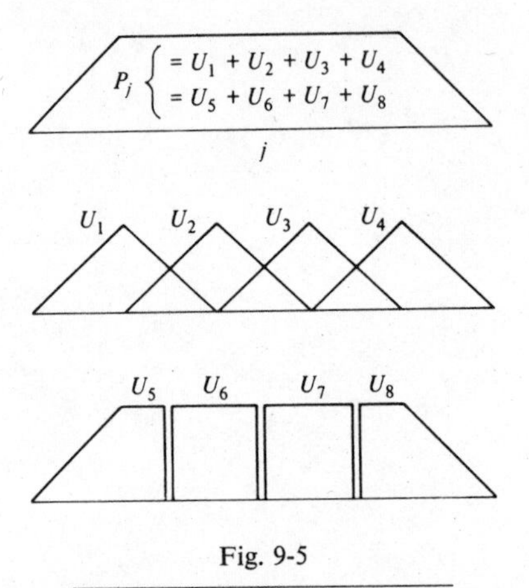

Fig. 9-5

Production and utilization curves as in Fig. 9-4 except that now we exhibit two alternate sets of species that utilize the same spectrum of resources, and yet each is resistant to invasion from the other region. See the text for details.

forests; so does the bird watcher in North America; and yet the New World warblers and flycatchers are unrelated to the Old World ones. The families have evolved in parallel to occupy similar environments. Yet few Old World warblers and no New World warbler or New or Old World flycatcher has ever really successfully invaded the other world. New World myrtle warblers (*Dendroica coronata*) enter slightly into Siberia and Old World Arctic warblers (*Phylloscopus borealis*) breed in northwestern Alaska, but each returns to winter with the rest of its species and neither has penetrated farther, to where there would be many competitors. We conclude that the difficulty of changing winter quarters, coupled with the diffuse competition from the other fauna, has prevented most exchanges of species. There has been some, but limited, exchange of less migratory species. Again, where the American bird species meet the tropical ones on the northeast coast of Mexico, there is a fairly abrupt transition. Few tropical species reach far beyond the Rio Grande, partly because they cannot stand the winters, and very few temperate species penetrate as far as Veracruz along the Mexican coast. In this case we must put most of the blame on diffuse competition.

This integrity of alternate communities provides the explanation for the patterns noticed by the earliest biogeographers. These men—Wallace and Sclater, for example—divided the world into biogeographic "realms," as in Fig. 9-6. The realms were Holarctic (divided into Nearctic and Paleartic), Neotropical, Ethiopian, Oriental, and Australian. The definition of realm was in part made possible by the independent faunas evolving in the separate areas and exchanging relatively few species. Their boundaries also match the great natural barriers to terrestrial exchange such as the Sahara Desert, the Himalayan plateau, and waterways separating continents.

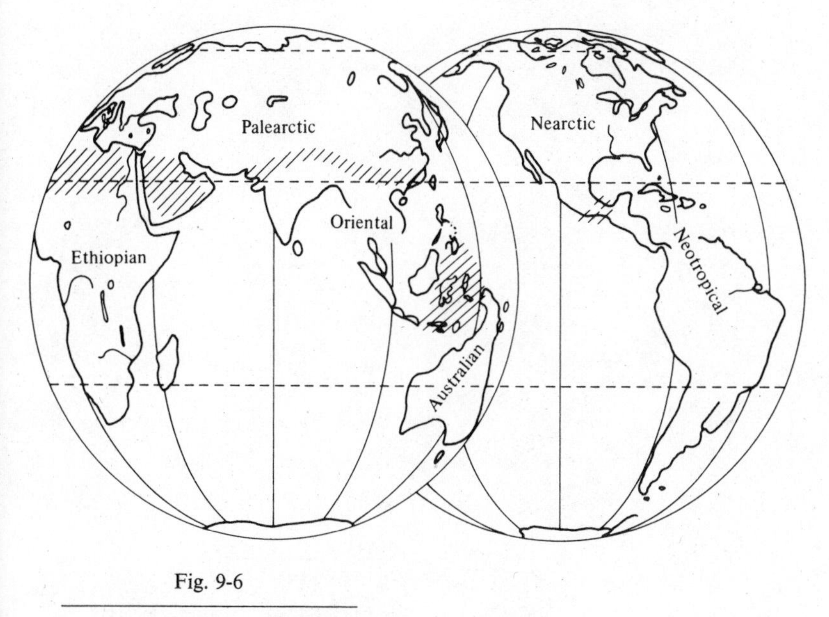

Fig. 9-6

The six continental faunal regions. Diagonal hatching shows approximate boundaries and transition areas. (From Darlington, 1957.)

Bibliography

Andrewartha, H. G., and Birch, L. C. 1954. The distribution and abundance of animals. Univ. Chicago Press, Chicago.

Ayala, F. 1971. Competition between species: frequency dependence. Science 171: 820–824.

Brodkorb, P. 1971. Origin and evolution of birds. *In* D. S. Farner and J. R. King [ed.] Avian biology. Academic, New York.

Brown, J. 1971a. Mechanisms of competitive exclusion between two species of chipmunks. Ecology 52: 305–311.

——— 1971b. Mammals on mountaintops: nonequilibrium insular biogeography. Amer. Natur. 105: 467–478.

Byers, H. 1954. The atmosphere up to 30 kilometers. *In* G. P. Kuiper [ed.] The earth as a planet. Univ. Chicago Press, Chicago.

Cody, M. L. 1966. A general theory of clutch size. Evolution 20: 174–184.

——— 1968. On the methods of resource division in grassland bird communities. Amer. Natur. 102: 107–147.

——— 1971. Finch flocks in the Mohave Desert. Theoret. Population Biol. 2: 142–158.

Cohen, D. 1967. Optimizing reproduction in a randomly varying environment. J. Theoret. Biol. 16: 1–14.

Crombie, A. C. 1946. Further experiments on insect competition. Roy. Soc. (London), Proc. B. 133: 76–109.

Crowell, K. L. 1962. Reduced interspecific competition among the birds of Bermuda. Ecology 43: 75–88.

Darlington, P. J. 1957. Zoogeography. Wiley, New York.

——— 1965. Biogeography of the southern end of the world. McGraw-Hill, New York.

Davis, Margaret B. 1969. Palynology and environmental history during the Quaternary Period. Amer. Sci. 57(3): 317–332.

Diamond, J. M. 1969. Avifaunal equilibria and species turnover rates on the Channel Islands of California. Nat. Acad. Sci., Proc. 64: 57–63.

Diamond, J. M. 1970a. Ecological consequences of island colonization by Southwest Pacific birds, I. Types of niche shifts. Nat. Acad. Sci., Proc. 67: 529–536.

——— 1970b. Ecological consequences of island colonization by Southwest Pacific birds, II. The effect of species diversity on total population density. Nat. Acad. Sci., Proc. 67: 1715–1721.

——— 1972. Comparison of faunal equilibrium turnover rates on a tropical and a temperate island. Nat. Acad. Sci., Proc. 68: 2742–2745

——— (In press.) The avifauna of the eastern highlands of New Guinea. Publ. Nuttall Ornith. Club, Cambridge, Mass.

Eisenmann, E. 1955. The species of Middle American birds. Linnaean Soc. N.Y., Trans. 7: 1–128.

Elton, C. 1958. The ecology of invasion by animals and plants. Methuen, London.

Farner, D. S. 1955. Birdbanding in the study of population dynamics. *In* A. Wolfson [ed.] Recent studies in avian biology. Univ. of Illinois Press, Urbana.

Fischer, A. 1960. Latitudinal variations in organic diversity. Evolution, 14(1): 64–81.

Flint, R. F. 1957. Glacial and Pleistocene geology. Wiley, New York.

Flohn, H. 1969. Climate and weather. World Univ. Library, McGraw-Hill, New York.

Franklin, J. N. 1968. Matrix theory. Prentice-Hall, Englewood Cliffs, N.J.

Fretwell, S. 1972. Populations in a seasonal environment. Princeton Univ. Press, Princeton, N.J.

Gause, G. F. 1934. The struggle for existence. Williams & Wilkins, Baltimore.

Grant, P. R. 1965. A systematic study of the territorial birds of the Tres Marias Islands, Mexico. Postilla 90.

Grinnell, J. 1914. An account of the mammals and birds of the lower Colorado Valley. Univ. Calif. Publ. Zool. 12: 51–294.

——— 1917. The niche-relationships of the California thrasher. Auk 34: 427–433.

——— 1943. Joseph Grinnell's philosophy of nature. Univ. of California Press, Berkeley.

Grinnell, J. and Orr, R. T. 1934. Systematic review of the *Californicus* group of the rodent genus *Peromyscus*. J. Mammal. 15: 210–220.

Haffer, J. 1969. Speciation in Amazonian forest birds. Science 165: 131–137.

Hall, E. R., and Kelson, K. R. 1959. The mammals of North America. Ronald, New York.

Hamilton, T. H., Barth, R. H., and Rubinoff, I. 1964. The environmental control of insular variation in bird species abundance. Nat. Acad. Sci., Proc. 52: 132–140.

Hastings, J. R., and Turner, R. M. 1965. The changing mile. An ecological study of vegetation change with time in the lower mile of an arid and semiarid region. Univ. of Arizona Press, Tucson.

Haverschmidt, F. 1968. Birds of Surinam. Oliver & Boyd, Edinburgh.

Hespenheide, H. 1971. Food preference and the extent of overlap in some insectivorous birds, with special reference to Tyrannidae. Ibis 113: 59–72.

Holdridge, L. R. 1967. Life zone ecology. Tropical Science Center, San José, Costa Rica.

Horn, H. S. 1966. Measurement of overlap in comparative ecological studies. Amer. Natur. 100: 419–424.

Huffaker, C. B. 1958. Experimental studies on predation. Hilgardia 27: 343–383.

Hutchinson, G. E. 1959. Homage to Santa Rosalia, or why are there so many kinds of animals. Amer. Natur. 93: 145–159.

Janzen, D. H. 1970. Herbivores and the number of tree species in tropical forests. Amer. Natur. 104: 501–529.

Jordan, Carl F. 1971. A world pattern in plant energetics. Amer. Sci. 59: 425–433.

Karr, James R. 1971. Structure of avian communities in selected Panama and Illinois habitats. Ecol. Monogr. 41(3): 207–233.

Kikkawa, J., and Williams, W. T. 1971. Altitudinal distribution of land birds in New Guinea. Search 2: 64–69.

Krebs, C., Keller, B., and Tamarin, R. 1969. *Microtus* population biology. Ecology 50: 587–607.

Kurten, B. 1969. Continental drift and evolution. Sci. Amer. 220: 54–65.

Lack, D. 1971. Ecological isolation in birds. Blackwell, Oxford.

Levins, R. 1968. Evolution in changing environments. Princeton Univ. Press, Princeton, N.J.

——— and Culver, D. 1971. Regional coexistence of species and competition between rare species. Nat. Acad. Sci., Proc. 68: 1246–1248.

Lowe, C. H., Heed, W. B., and Halpern, E. A. 1967. Supercooling of the saguaro species *Drosophila nigrospiracula* in the Sonoran Desert. Ecology 48: 984–985.

MacArthur, R. 1959. On the breeding distribution pattern of North American migrant birds. Auk 76: 318–325.

——— 1965. Patterns of species diversity. Biol. Rev. 40: 510–533.

——— 1968. The theory of the niche. Population biology and evolution. Syracuse Univ. Press, Syracuse, N.Y.

——— 1969. Patterns of communities in the tropics. Biol. J. Linnaean Soc. 1: 19–30.

——— 1970. Species packing and competitive equilibrium among many species. Theoret. Population Biol. 1: 1–11.

——— and MacArthur, J. 1961. On bird species diversity. Ecology 42: 594–598.

——— and Connell, J. 1966. The biology of populations. Wiley, New York.

——— and Pianka, E. 1966. On optimal use of a patchy environment. Amer. Natur. 100: 603–609.

——— and Wilson, E. O. 1967. The theory of island biogeography. Princeton Univ. Press, Princeton, N.J.

——— Recher, H., and Cody, M. 1966. On the relation between habitat selection and species diversity. Amer. Natur. 100: 319–332.

——— Diamond, J., and Karr, J. (In press.) Density compensation in island faunas. Submitted to *Ecology*.

Maly, E. 1969. A laboratory study of the interaction between the predatory rotifer *Asplanchna* and *Paramecium*. Ecology 50: 59–73.

Martin, P. 1971. Ecology in the fourth dimension: Population extinctions during the Pleistocene. *In* Topics in the study of life. Harper & Row, New York.

Mayr, E., and L. L. Short, 1970. Species taxa of North American Birds. Publ. Nuttall Ornith. Club 9, Cambridge, Mass.

Merriam, C. H. 1890. Results of a biological survey of the San Francisco Mountain region and desert of the Little Colorado in Arizona. U.S. Dep. Agr., N. Amer. Fauna 3: 1–136.

Moll, E. O., and Legler, J. M. 1971. The life history of a neotropical slider turtle, *Pseudomys scripta* (Schoepff), in Panama. Bull. Los Angeles County Mus. Nat. Hist. Sci. 11.

Mooney, H. A., and Billings, W. D. 1961. Comparative physiological ecology of arctic and alpine populations of *Oxyria digyna*. Ecol. Mongr. 31: 1–29.

Niering, W. A., Whittaker, R. H., and Lowe, C. H. 1963. The saguaro: a population in relation to environment. Science 142: 15–23.

Nisbet, I. C. T. 1971. The laughing gull in the Northeast. Amer. Birds 25: 677–683.

Odum, Howard T., and Pinkerton, Richard C. 1955. Time's speed regulator. Amer. Sci. 43(2): 331–343.

Paine, R. T. 1966. Food web complexity and species diversity. Amer. Natur. 100: 65–76.

Park, T. 1962. Beetles, competition and populations. Science 138: 1369–1375.

Patrick, R. 1963. The structure of diatom communities under varying ecological conditions. Ann. N.Y. Acad. Sci. 108: 353–358.

——— 1966. The Catherwood Foundation Peruvian Amazon Expedition. Monographs of Academy of Natural Sciences of Philadelphia.

——— 1968. The structure of diatom communities in similar ecological conditions. Amer. Natur. 102: 173–184.

——— 1970. Benthic stream communities. Amer. Sci. 58: 546–549.

——— Hohn, M., and Wallace, J. 1954. A new method of determining the pattern of the diatom flora. Notulae Natura 259. Academy of Natural Sciences of Philadelphia.

Pearsall, W. H. 1950. Mountains and moorlands. Collins, London.

Peterson, R. T. 1961. A field guide to the western birds. Houghton Mifflin, Boston.

——— R. T. 1963. A field guide to the birds of Texas. Houghton Mifflin, Boston.

Pianka, E. 1969. Habitat specificity, speciation and species density in Australian desert lizards. Ecology 50: 498–502.

Recher, H. F. 1969. Bird species diversity and habitat diversity in Australia and North America. Amer. Natur. 103: 75–80.

Robbins, C. S., Bruun, B., and Zim, H. S. 1966. Birds of North America. Golden Press, New York.

——— and Van Velzen, W. T. 1969. Breeding bird survey 1967 and 1968. Bureau of Sport Fisheries and Wildlife Special Scientific Report 124. Washington, D.C.

Rosenzweig, M., and MacArthur, R. 1963. Graphical representation and stability conditions of predator prey interactions. Amer. Natur. 97: 209–223.

Roughgarden, J. 1971. Density dependent natural selection. Ecology 52: 453–468.

Salisbury, E. J. 1942. The reproductive capacity of plants. Bell, London.

Sanders, H. L. 1968. Marine benthic diversity: a comparative study. Amer. Natur. 102: 243–282.

——— 1969. Benthic marine diversity and the stability time hypothesis. *In* Diversity and stability in ecological systems. Brookhaven Symp. Biol. 22.

Schauensee, R. M. 1964. The birds of Colombia. Livingston Publishing Company, Wynnewood, Pa.

Schoener, T. W. 1967. The ecological significance of sexual dimorphism in size in the lizard *Anolis conspersus*. Science 155: 474–477.

——— 1969. Optimal size and specialization in constant and fluctuating environments: An energy-time approach. *In* Diversity and stability in ecological systems. Brookhaven Symp. Biol. 22.

——— 1971. Large-billed insectivorous birds: a precipitous diversity gradient. Condor 73: 154–161.

——— and Janzen, D. 1968. Notes on environmental determinants of tropical versus temperate insect size patterns. Amer. Natur. 102: 207–224.

Simberloff, D. S., and Wilson, E. O. 1969. Experimental zoogeography of islands. The colonization of empty islands. Ecology 50: 278–296.

——— and Wilson, E. O. 1970. Experimental zoogeography of islands. A two-year record of colonization. Ecology 51: 934–937.

Simpson, G. G. 1965. The geography of evolution. Chilton, Philadelphia.

Skellam, J. G. 1951. Random dispersal in theoretical populations. Biometrika 38: 196–218.

Skutch, A. F. 1954. Life histories of Central American birds I. Pacific Coast Avifauna 31. Cooper Ornith. Soc.

——— 1960. Life histories of Central American birds II. Pacific Coast Avifauna 34. Cooper Ornith. Soc.

——— 1967. Life histories of Central American highland birds. Publ. Nuttall Ornith. Club 7, Cambridge, Mass.

——— 1969. Life histories of Central American birds III. Pacific Coast Avifauna 35. Cooper Ornith. Soc.

Slobodkin, L., and Sanders, H. 1969. On the contribution of environmental predictability to species diversity. *In* Diversity and stability in ecological systems. Brookhaven Symp. Biol. 22.

Smith, C. 1970. The coevolution of pine squirrels (*Tamiasciurus*) and conifers. Ecol. Monogr. 40(3): 349–371.

Smythe, N. 1970. Neotropical fruiting seasons and seed dispersal. Amer. Natur. 104: 25–35.

Snow, D. W. 1962. A field study of the black and white manakin (*Manacus manacus*) in Trinidad. Zoologica 47: 65–104.

Stehli, F. G. 1968. Taxonomic diversity gradients in pole location: the recent model. *In* Evolution and environment. Peabody Museum Centennial Symposium, Yale Univ. Press, New Haven, Conn.

Storer, R. W. 1966. Sexual dimorphism and food habits in three North American accipiters. Auk 83: 423–436.

Terborgh, J. 1971. Distribution on environmental gradients: theory and a preliminary interpretation of distributional patterns in the avifauna of the Cordillera Vilcabamba, Peru. Ecology 52: 23–40.

Thom, R. 1970. Topological models in biology. *In* C. H. Waddington [ed.] Towards a theoretical biology. Aldine, Chicago.

United States Department of Agriculture. 1941. Climate and man. Washington, D.C.

United States Department of Agriculture. 1965. Silvics of the forest trees of the United States. Washington, D.C.

Vandermeer, J. 1970. The community matrix and the number of species in a community. Amer. Natur. 104: 73–84.

Vanzolini, P. E., and Williams, E. E. 1970. South American Anoles: the geographic differentiation and evolution of the *Anolis chrysolepis* species group (Sauria, Iguanidae) Arq. Zool. 19: 1–124.

Vuilleumier, F. 1970. Insular biogeography in continental regions. The northern Andes of South America. Amer. Natur. 104: 373–388.

Webb, S. D. 1969. Extinction–origination equilibria in late cenozoic land mammals of North America. Evolution 23: 688–702.

Wetmore, A. 1957. The birds of Isla Coiba, Panama. Smithsonian Inst. Misc. Collection 134(9): 1–105.

Whittaker, R. H. 1969. Evolution of diversity in plant communities. *In* Diversity and stability in ecological systems. Brookhaven Symp. Biol. 22.

Williams, E. E. 1969. The ecology of colonization as seen in the zoogeography of Anoline lizards on small islands. Quart. Rev. Biol. 44: 345–389.

Wilson, E. O. 1958. Patchy distribution of ant species in New Guinea rain forests. Psyche 65: 26–38.

——— 1959. Adaptive shift and dispersal in a tropical ant fauna. Evolution 13: 122–144.

——— 1961. The nature of the taxon cycle in the Melanesian ant fauna. Amer. Natur. 95: 169–193.

——— 1969. The species equilibrium. *In* Diversity and stability in ecological systems. Brookhaven Symp. Biol. 22.

——— 1971. Competitive and aggressive behavior. *In* J. F. Eisenberg and W. Dillon [ed.] Man and beast; comparative social behavior. Smithsonian Institution Press, Washington, D.C.

——— and Simberloff, D. S. 1969. Experimental zoogeography of islands. Defaunation and monitoring techniques. Ecology 50: 267–278.

——— and Taylor, R. W. 1967. An estimate of the potential evolutionary increase in species density in the Polynesian ant fauna. Evolution 21: 1–10.

Index

References to illustrations are printed in italic type.

Library of Congress Cataloging in Publication Data

MacArthur, Robert H.
Geographical ecology.

Reprint. Originally published: New York:
Harper & Row, 1972.
Bibliography: p.
Includes index.
1. Biogeography. 2. Ecology. I. Title.
QH84.M23 1984 574.5'24 83-24477
ISBN 0-691-08353-3 (alk. paper)
ISBN 0-691-02382-4 (pbk.)